CW01494355

AUSTIN/MG

MAESTRO
MONTEGO
1983 to 1992

All Austin/MG Maestro and Montego models (including Vanden Plas),
except Maestro Van and Diesel models.

A Haynes Handbook
and Drivers Guide
© Haynes Publishing Group 1992

Printed and published by
J H Haynes & Co Ltd Sparkford
Nr Yeovil Somerset BA22 7JJ
England

ISBN 1 85010 819 6

**British Library Cataloguing in
Publication Data.** A catalogue
record for this book is available
from the British Library

HANDBOOK & DRIVERS GUIDE
by Jeremy Churchill
Full details of all servicing and repair tasks for the models covered by this
Handbook can be found in the relevant Owners Workshop Manual –
OWM **922** Maestro 1.3 & 1.6, OWM **1066** Montego 1.3 & 1.6, or
OWM **1067** Montego 2.0.

ABCDEFGHIJKLMNOPQRST

2 ACKNOWLEDGEMENTS

We gratefully acknowledge the assistance of RoSPA in compiling the information used in this Handbook. Thanks are also due to Duckhams Oils who provided lubrication data, and to Britax who supplied information on child safety. Additional photographs were supplied by Quadrant Picture Library/Auto Express and David Sparrow.

Thanks are also due to all those people at Sparkford who helped in the production of this Handbook.

We take great pride in the accuracy of information given in this Handbook, but vehicle manufacturers make alterations and design changes during the production run of a particular vehicle of which they do not inform us. No liability can be accepted by the authors or publishers for loss, damage or injury caused by any errors in, or omissions from, the information given.

AUSTIN/MG MAESTRO & MONTEGO

The idea behind this Handbook is to help you to get the most out of your motoring.

Apart from the things that every owner needs to know, to deal with unexpected mishaps like a puncture or a blown light bulb, you'll find clearly-presented information on road safety, hints on driving abroad, and tips on how to prepare your car for the MOT test. We've also included details of local radio frequencies to help you avoid those inevitable 'jams', and there's a Section on what to do in the unfortunate event of an accident.

For those not familiar with their car, there's a Section to explain the location and the operation of the various controls and instruments. Additionally, there's a Section on fault finding, and a useful glossary of car jargon.

Garage labour charges usually form the major part of any car servicing bill, and we hope to help you to reduce those bills by carrying out the more straightforward routine servicing jobs yourself. If you're about to start carrying out your own servicing for the first time, we aim to provide you with easy-to-follow instructions, enabling you to carry out the simpler tasks, which perhaps you've left to a garage or a 'car-minded' friend in the past. Even if you prefer to leave regular servicing to a suitably-qualified expert, by using this book you'll be able to carry out regular checks on your car, to make sure that your motoring is safe and hopefully trouble-free. You'll also find advice on buying suitable tools, and safety in the home workshop.

Some readers of this Handbook may not yet have bought a Maestro or Montego, so we've included a brief history of the range, and some useful tips on buying and selling.

All in all, we hope that this book will prove a handy companion for your motoring adventures, and hopefully we'll help to reduce the problems which inevitably crop up in everyday driving. If you're bitten by the DIY bug, and you're keen to tackle some of the more advanced repair jobs on your car, then you'll need our **Owners Workshop Manual** for your particular model (OWM 922 for Maestro 1.3 & 1.6 models, OWM 1066 for Montego 1.3 & 1.6 models or OWM 1067 for Montego 2.0 models). These manuals give a step-by-step guide to all the repair and overhaul tasks, with plenty of illustrations to make things even clearer.

Happy motoring!

ABOUT THIS HANDBOOK

▲ Maestro 1.3 Special (1989)

▲ Maestro 1.6 HL (1986)

The Maestro, launched in March 1983, is available only in a five-door Hatchback bodystyle; the Montego was launched a year later, first in four-door Saloon, then in five-door Estate bodystyles. Both models share the maximum possible number of components, and arrived with a wide range of standard and optional equipment; the range has been updated subsequently to maintain equipment levels on a par with the competition.

1.3 litre models are fitted with the 1275 cc 'A+' overhead-valve engine, while early Maestro 1.6 litre models have the 1598 cc 'R-series' overhead-camshaft engine. Montego 1.6 litre models (and all Maestro 1.6 litre models from May 1984) use the 1598 cc 'S-series' overhead-camshaft engine, while all 2.0 litre models are fitted with the 1994 cc 'O-series' overhead-camshaft engine.

All gearboxes are bolted to the left-hand end of the transversely-mounted engine; 1.3 & (early Montego) 1.6 litre models are fitted with Volkswagen-built four- or five-speed manual units, with a Volkswagen three-speed automatic transmission optional on some 1.6 litre models. 1.6 litre Montegos from late 1988 onwards are fitted with a Honda-designed, Austin Rover-built, five-speed manual gearbox. All 2.0 litre models are fitted with a Honda-designed five-speed manual gearbox, and a ZF four-speed automatic transmission was optional on all 2.0 litre Montegos (except MGs) from April 1986.

Fuel and ignition systems are equally diverse; apart from the Maestro City models (which have conventional SU carburettors with manual chokes), all 1.3 litre models employ electronically-controlled SU carburettors and Lucas electronic ignition. Apart from the MG Maestro 1600 models (which have twin Weber carburettors), all 1.6 litre models, 2.0 litre carburettor-engined models & MG Turbo models use electronically-controlled SU carburettors and Lucas programmed electronic ignition. On Montegos from late 1988 onwards, both ignition and carburettor are linked by a single engine management control unit. 2.0 litre fuel injection-engined models are fitted with Lucas EFi (Electronic Fuel injection), except for some later Montego models, which may be fitted with Rover/Motorola MEMS-MPi (Modular Engine Management System – Multi-Point injection).

While these cars were designed to require as little regular servicing as possible, several operations can be undertaken only by the well-equipped DIY mechanic, and others are only for a Rover dealer.

▲ *Montego 2.0 LX (1990)*

THE MAESTRO & MONTEGO

▲ *Montego 1.6 Countryman Estate (1989)*

▲ *MG Montego Turbo 2.0 (1986)*

GENERAL TIPS ON BUYING

- Don't rush out and buy the first car to catch your attention
- Always buy from a recognised dealer in preference to buying privately
- Have the car checked over by someone knowledgeable before you buy it
- In general, you get what you pay for!

Before buying a second-hand car, it's worthwhile doing some homework to try and avoid some of the pitfalls waiting for the unwary. First of all, don't rush out and buy the first car to catch your attention (all that glitters is not gold!), and remember that much of the responsibility is yours when it comes to the soundness of the deal, especially when buying a car privately.

Wherever possible, buy from a recognised dealer, and check that the dealer is a member of the Retail Motor Industry Federation, as this will provide you with certain legal safeguards if you have any problems. If you buy from a dealer, you are covered by the Sale Of Goods Act, which in summary states that the goods must be fit for their intended purpose, the goods must be of proper quality, and the goods must be as described by the seller. If you're buying a car privately, ask to see the service receipts and the MOT certificates going back as long as the car has been in the possession of the current owner (this will help to establish that the car hasn't been stolen, and that the recorded mileage is genuine), and always ask to view the car at the seller's private address (to make sure that the car isn't being sold by an unscrupulous dealer posing as a private seller). Check the vehicle documents for obvious signs of forgery and, if in doubt, contact the DVLA and give them details of the Registration Document, as they will be able to run a check on its authenticity.

As far as the soundness of the car itself is concerned, a genuine service history is helpful. This is provided by the service book supplied with the car when new, which should be completed and officially stamped by an authorised garage after each service. Cars with a full service history (fsh) usually command a higher price than those without.

To check the condition of a car, a professional examination is well worthwhile, if you can afford it, and organisations such as the AA and RAC will be able to provide such a service. Otherwise, you must trust your own judgement, and/or that of a knowledgeable friend. Although it's tempting, try not to overlook the mechanical soundness of the car in favour of the overall appearance. It's relatively easy to clean and polish a car every week, but when were the brakes and tyres last checked? Above all, safety must always take priority. Don't view a car in the wet, as water on the bodywork can give a misleading impression of the condition of the paintwork. First of all, check around the outside of the car for rust, and for obvious signs of new or mismatched paintwork which might show that the car has been involved in an accident. Check the tyres for signs of unusual wear or damage, and check that the car 'sits' evenly on its suspension, with all four corners at a similar height. Open the bonnet and check for any obvious signs of fluid leakage (oil, water, brake fluid), then start the engine and listen for any unusual noises – some background noise is to be expected on older cars, but there should be no sinister rattles or bangs! Also listen to the exhaust to make sure that it isn't 'blowing' indicating the need for renewal, and check for signs of excessive exhaust smoke. Black smoke may be caused by poorly adjusted fuel mixture, which can usually be rectified fairly easily, but blue smoke usually indicates worn engine components, which may prove expensive to repair. Finally, drive the car, and test the brakes, steering and gearbox. Make sure that the car doesn't pull to one side, and check that the steering feels positive and that the gears can be selected satisfactorily without undue harshness or noise. Listen for any unusual noises or vibration, and keep an eye on the instruments and warning lights to make sure that they are working and indicating correctly.

If all proves satisfactory, try to negotiate a suitable deal, but remember that a dealer has to work to a profit margin, and it's unlikely that you'll find a good car for a silly price. Always obtain a receipt for your money.

On the whole, it's true to say that you get what you pay for. Above all, don't be rushed into making a hasty decision.

POINTS TO LOOK FOR WHEN BUYING A MAESTRO/MONTEGO

Early Maestro/Montegos gained a (in many cases, well-justified) reputation for poor assembly and reliability, and should be avoided; while problems caused by poor assembly should have been cured by now on a car that is used regularly, they can still be unreliable, and their awful reputation means that such a car may be worth much less than you expect if you should decide to resell. This means avoiding Maestros built before 1985 ('C' registration), particularly the MG 1600 model with its troublesome twin-Weber carburettor engine, and Montegos built before 1987 ('E' registration). Note also that Montego 1.3 models are slow, and less economical than their 1.6 litre counterparts.

Later models (as recent as you can afford) on the other hand, are well-equipped and offer economical and reliable motoring. If you buy at a price that takes advantage of their poor reputation, and do not resell for some time, they represent excellent value for money. Remember, though, that any MG model will be in much higher insurance groups, and that MG Turbo models should be avoided, unless you are very sure of the car's past use and service history, and don't mind the high running costs.

First examine the car's interior; if you are looking at one of the upmarket models with electric windows, central locking and/or the solid-state instruments/voice synthesis unit, check that all these systems are working properly, as they are very expensive to repair. Check for damp patches (or chalky white deposits) on the trim components around the windscreen and rear window; if these or any other signs of water leaks are found, reject the car, as such leaks are almost impossible to cure.

On the outside, check for patches of rust and for poorly-fitting doors or other signs of badly-repaired accident damage. Rust should not be a problem on any of these cars; the only areas that require a careful check are the sill panels (under the doors), the vertical seam at each end of each sill panel, and the bottom edges of the doors themselves. Also check the bottom edge of the tailgate on Maestros, and the panel between the bottom of the rear window and the edge of the boot opening on Montego Saloons. If significant rust or bubbling paintwork is found on any Maestro/Montego, reject it, as there will be plenty of rust-free models around. Any unsightly surface rust due to neglected stone chips on the bonnet and around the wheel arches, can be used to reduce the asking price if you feel like bargaining, but remember that bodywork repairs can be surprisingly expensive.

If either bumper shows signs of damage such as cracks or flaking paintwork, or if either is badly-fitted, first check behind the bumper for accident damage; if no body damage is found, negotiate a reduction in the asking price, remembering that these bumpers are very expensive.

Moving to the engine compartment, check for signs of oil leaks; be suspicious if the engine has been very recently steam-cleaned, or if some parts of the engine/transmission appear 'cleaner' than others. If the engine has been warmed-up for your inspection, the seller may be hiding something.

Particularly on 1.6 & 2.0 litre models, note the coolant level in the expansion tank, and check for signs of water droplets in the oil on the dipstick; the oil itself should of course be clean, and light gold in colour. Remove the oil filler cap, and check for 'mayonnaise' (a creamy, yellowish-grey substance with traces of water droplets in it) on the underside of the filler cap and inside the engine/oil filler tube. Small amounts of this may merely indicate that the car has only been used on short trips recently. Any significant deposits (which seem to be an inherent fault on some models, leading to engine damage because of blocked oilways) may mean that the cylinder head gasket is leaking, allowing the coolant to contaminate the engine oil. This is expensive to have repaired, so look carefully for other symptoms of this fault.

On 1.6 & 2.0 litre models which still have the electronically-controlled carburettor (there are several good manual-choke aftermarket conversions around) and fuel-injected 2.0 litre models, ask the seller to start the engine from stone cold and listen carefully; the engine should start immediately and run smoothly at a high idle speed, then smoothly and steadily slow to normal idle as it warms up. If there is any sign of hesitancy, roughness or misfiring, or if the engine actually cuts out on idling, reject the car; attention either to electronically-controlled carburettors or to the fuel injection

system is expensive and not always successful.

When taking 1.6 or 2.0 litre models for a test drive, watch the temperature gauge for overheating; this may be a further sign that the head gasket is faulty.

On 1.3 litre models, difficult starting and poor idling can usually be cured by careful (but probably expensive) servicing. On the test drive, change up to 3rd gear from a steady 15 mph on a flat road and accelerate; if a deep rumble or knock suddenly becomes more obvious, reject the car, especially if it has covered around 60 000 miles – the main and/or big-end bearings are probably worn. When the test drive is over and the engine is fully warmed-up, check the exhaust; if it shows traces of blue smoke, the engine is well-worn, and the car should be rejected.

On all models, listen for any strange noises; if a constant low rumble is heard from the front, the wheel bearings may be worn, which usually requires the renewal of the bearing(s) on both sides to cure the problem. Montegos seem to be prone to rapid front brake pad and disc wear; if a loud squealing is heard when the brakes are applied, it may be just the pads (cheap enough), but if the noise is accompanied by vibration felt through the brake pedal or steering, the discs may be due for renewal (expensive).

On 1.6 & 2.0 litre models, when the test drive is complete, open the bonnet and check the level of coolant in the expansion tank; if it has dropped significantly this, in addition to the points previously noted, indicates a head gasket fault, and the car should be rejected. Also on these engines, listen carefully to the fully warmed-up engine ticking over; a light, regular rattle should be clearly audible from the top end, which gets louder and quicker as engine speed is increased. If the engine is quiet or the noise is irregular, reject the car as the valve clearances are probably too tight (showing that important servicing has been neglected or badly-performed) and the valve seats may have been damaged. On models so equipped, check that the power steering pump is silent when the engine is idling; if it whines (especially when the steering is turned to full lock) it is badly-worn, and will be expensive to renew.

GENERAL TIPS ON SELLING

● **Make sure that the car is clean and tidy**
● **Make sure that all of the service documents, registration document etc, are available for inspection**
● **Ask yourself ... 'Would I buy this car?'**

Obviously when selling a car, bear in mind the points which the prospective buyer should be looking for, as described in the above sections. It goes without saying that the car should be clean and tidy, as first impressions are important. Any fluid leaks should be cured, and there's no point in trying to disguise any major bodywork or mechanical problems.

If you're trading the car in with a dealer, you will always get a lower price than if you sell privately, but you can be fairly sure that there will be less comeback to you should any unexpected problems develop. If selling privately, don't allow the buyer to take the car away until you have his/her money, and it's a good idea to ask him/her to sign a piece of paper to say that he/she is happy to buy the car as viewed, just in case any problems develop later on. Give a receipt for the money paid.

Note: All figures are approximate, and will vary depending on model

DIMENSIONS [mm (in)]

Overall length

Maestro models	**3998** to **4079** (157.5 to 160.7)
Montego models	**4465** to **4468** (175.9 to 176.0)

Overall width

All models – including mirrors	**1941** (76.5)

Overall height

Maestro models	**1412** to **1435** (55.6 to 56.5)
Montego Saloon models	**1402** to **1420** (55.2 to 55.9)
Montego Estate models – including integral roof rack, where fitted	**1445** to **1522** (56.9 to 60.0)

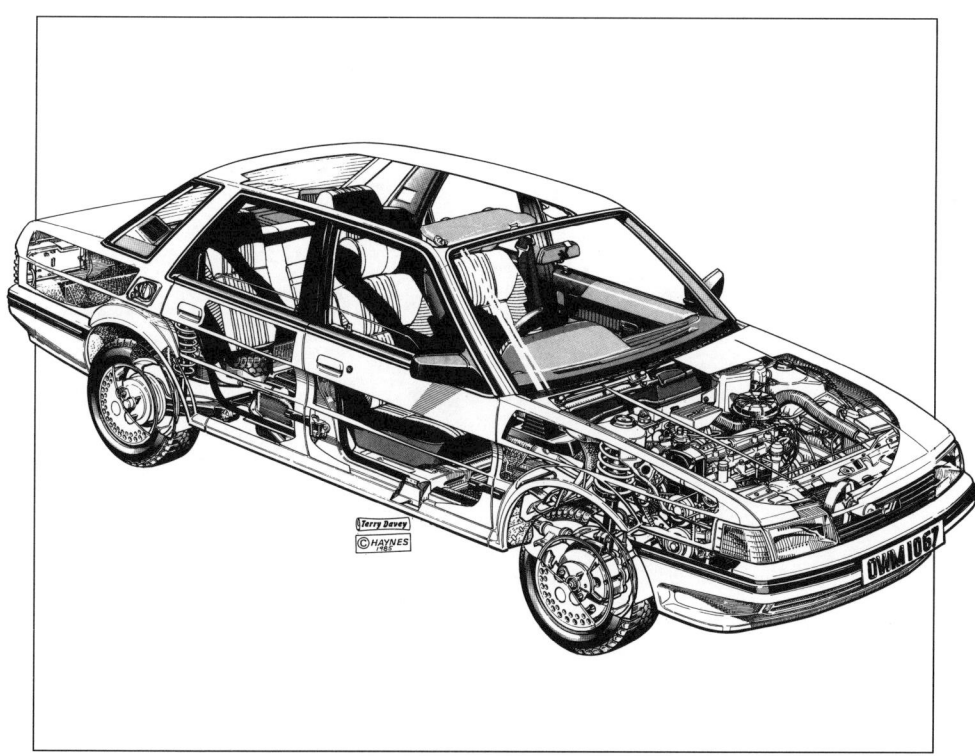

AUSTIN/MG MAESTRO & MONTEGO

Note: All figures are approximate, and will vary depending on model

WEIGHTS [kg (lb)]

Nominal kerb weight

Maestro models	**875** to **1071** (1929 to 2362)
Montego Saloon models	**949** to **1185** (2093 to 2613)
Montego Estate models	**1018** to **1206** (2245 to 2659)

Maximum roof rack load

Maestro models up to 1989	**75** (165)
Later Maestro models, Montego Saloon models, Montego Estate models with removable (accessory) roof rack	**70** (154)
Montego Estate models with integral roof rack *	**100** (221)

* *Refer to 'Controls and equipment' on page 38 for full details.*

Maximum towing weight (for trailer or caravan)

Maestro 1.3 litre models with '3 + E' or '4 + E' manual gearbox	**900** (1985)
All other Maestro 1.3 litre models, Maestro 1.6 litre automatic transmission models	**1000** (2205)
Maestro 1.6 litre manual gearbox models	**1125** (2481)
Maestro 2.0 litre models	**1146** (2527)
Montego 1.3 litre models	**950** (2095)
Montego 1.6 litre models	**1070** (2359)
Montego 2.0 litre models except those below	**1220** (2690)
MG Montego models, 1986-on	**1250** (2756)

Maximum trailer/caravan noseweight

All models	**34** to **45** (75 to 99)

For those not familiar with Maestro and Montego models, this Section will help to identify the instruments and controls. Typical instrument panel layouts are shown in the accompanying illustrations. The operation of most equipment is self-explanatory, but some items require further explanation to ensure that their use is fully understood.

Note that not all items are fitted to all models.

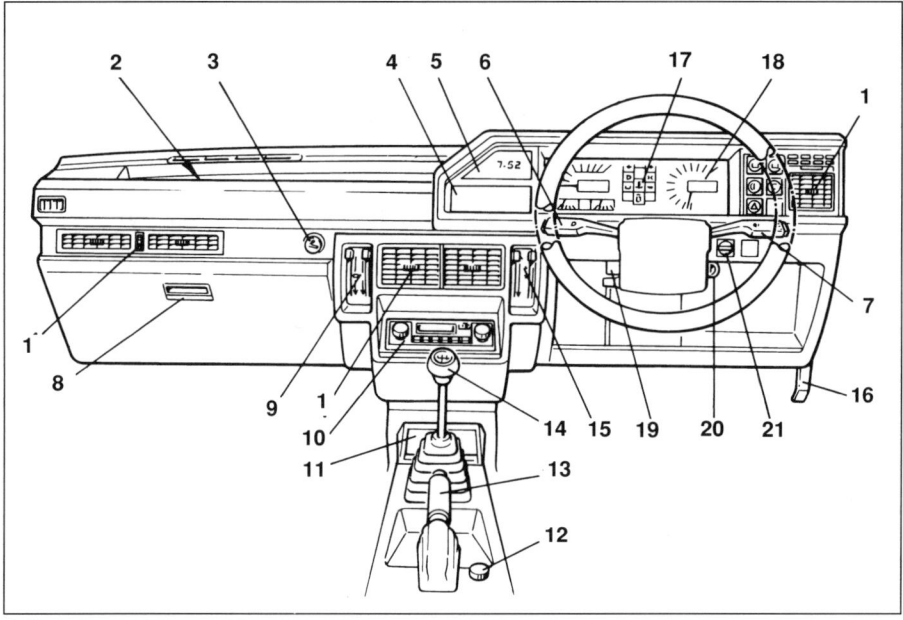

▲ Typical control layout – Maestro

1	Face level ventilation vents	11	Coin tray
2	Fusebox cover – early models	12	Loudspeaker fader control
3	Cigar lighter	13	Handbrake
4	Ashtray	14	Gear lever
5	Clock	15	Air distribution/blower controls
6	Headlamp dip/flash, horn and direction indicator switch	16	Bonnet release handle
7	Windscreen wash/wipe switch	17	Warning lamp panel
8	Glovebox release button	18	Instrument panel
9	Vent and temperature controls	19	Lighting switch
10	Radio/cassette player controls	20	Ignition switch/steering lock
		21	Instrument illumination rheostat

CONTROLS & EQUIPMENT

AUSTIN/MG MAESTRO & MONTEGO

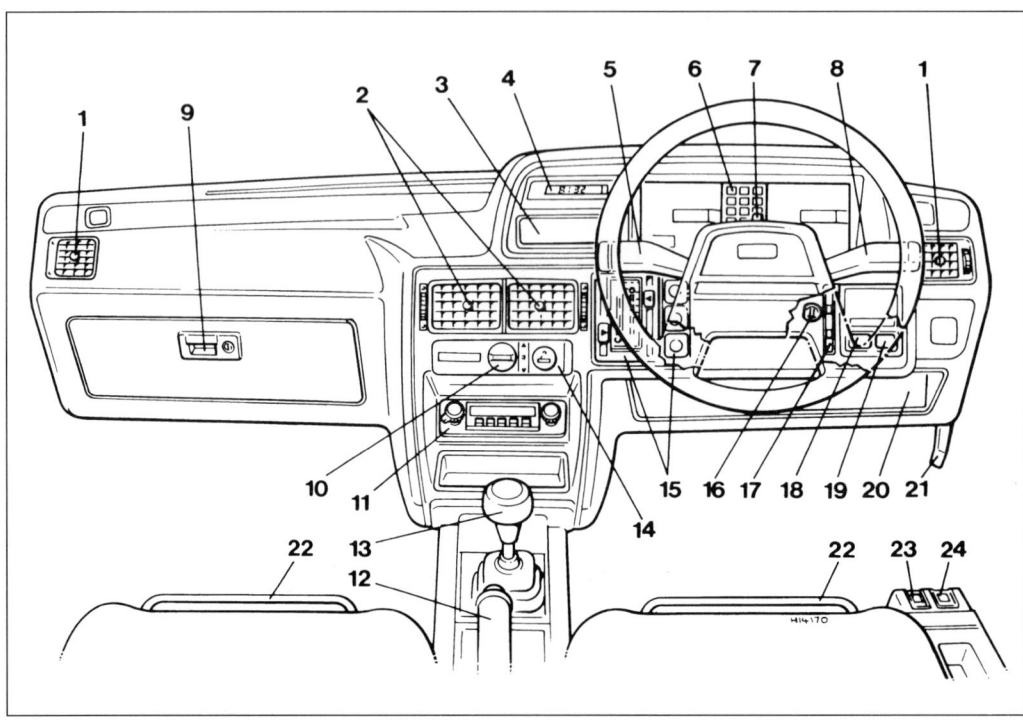

▲ *Typical control layout – Montego*

1   Fresh air vents
2   Central air vents
3   Ashtray
4   Clock
5   Lighting, headlamp dip/flash, horn and direction indicator switch
6   Instrument panel
7   Hazard warning switch
8   Windscreen wash/wipe and headlamp washer switch
9   Glovebox release
10   Loudspeaker fader control
11   Radio/cassette player controls
12   Handbrake

13   Gear lever
14   Cigar lighter
15   Heater controls
16   Ignition switch/steering lock
17   Instrument illumination rheostat
18   Rear foglamp switch
19   Heated rear window switch
20   Fusebox cover
21   Bonnet release handle
22   Seat release bar
23   Fuel filler flap release
24   Boot lid/tailgate release

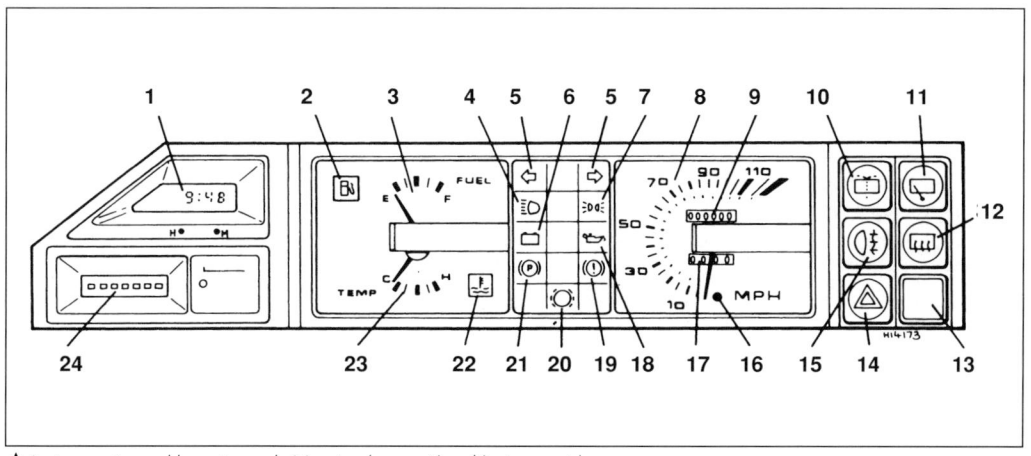

▲ *Instrument panel layout – early Maestro (conventional instruments)*

1	Clock	**13**	Spare switch position
2	Low fuel level warning lamp	**14**	Hazard warning switch
3	Fuel gauge	**15**	Rear foglamp switch
4	Main beam 'ON' warning lamp	**16**	Trip meter reset button
5	Direction indicator/hazard warning lamps	**17**	Trip mileage meter
6	Ignition/no-charge warning lamp	**18**	Low oil pressure warning lamp
7	Sidelamps 'ON' warning lamp	**19**	Low brake fluid level warning lamp
8	Speedometer	**20**	Brake pad wear warning lamp
9	Total mileage recorder	**21**	Handbrake 'ON' warning lamp
10	Rear window washer switch	**22**	High coolant temperature warning lamp
11	Rear window wiper switch	**23**	Temperature gauge
12	Heated rear window switch	**24**	Econometer

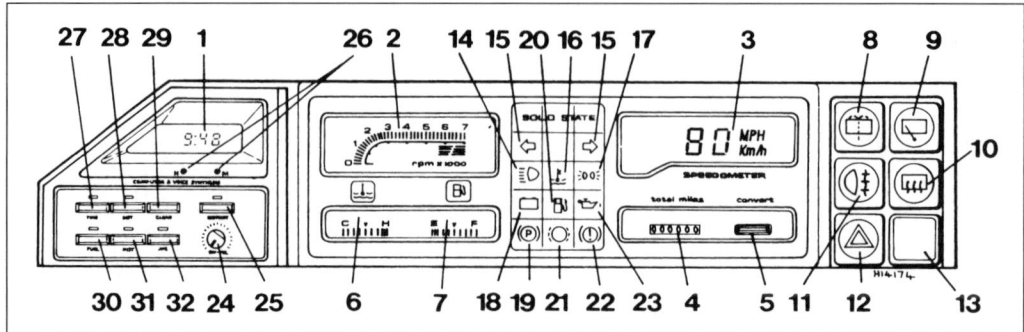

▲ *Instrument panel layout – Maestro (solid-state instruments)*

1 Clock/trip computer display
2 Tachometer ('rev. counter')
3 Speedometer
4 Total mileage recorder
5 Imperial/metric conversion button
6 Temperature gauge
7 Fuel gauge
8 Rear window washer switch
9 Rear window wiper switch
10 Heated rear window switch
11 Rear foglamp switch
12 Hazard warning switch
13 Spare switch position
14 Main beam 'ON' warning lamp
15 Direction indicator/hazard warning lamps
16 High coolant temperature warning lamp

17 Sidelamps 'ON' warning lamp
18 Ignition/no-charge warning lamp
19 Handbrake 'ON' warning lamp
20 Low fuel level warning lamp
21 Brake pad wear warning lamp
22 Low brake fluid level warning lamp
23 Low oil pressure warning lamp
24 Voice unit on/off and volume control
25 Voice unit 'Restrict' button
26 Time/date setting buttons
27 Time/date display button
28 Trip distance button
29 Trip computer 'Clear' button
30 'Fuel used' button
31 'Instantaneous fuel consumption' button
32 Average fuel consumption' button

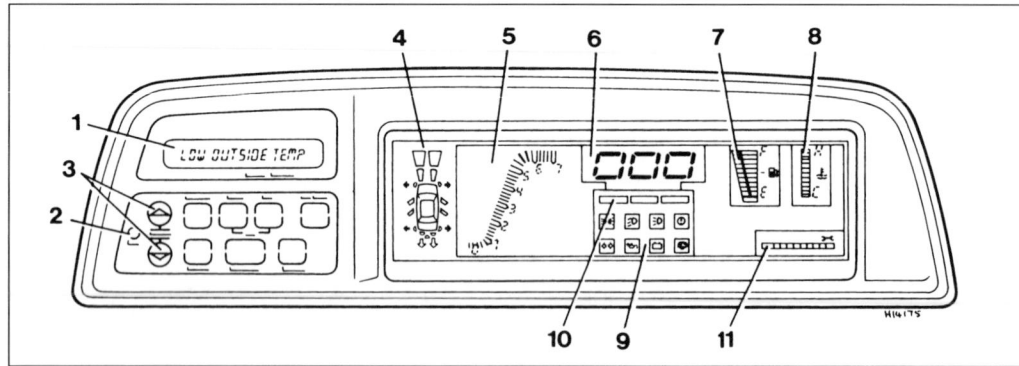

▲ *Instrument panel layout – Montego (Driver Information Centre)*

1 Message centre
2 Voice unit on/off switch
3 Voice unit volume controls
4 Vehicle map
5 Tachometer ('rev. counter')
6 Speedometer

7 Fuel gauge
8 Temperature gauge
9 Warning lamp panel
10 Attention display
11 Service interval indicator

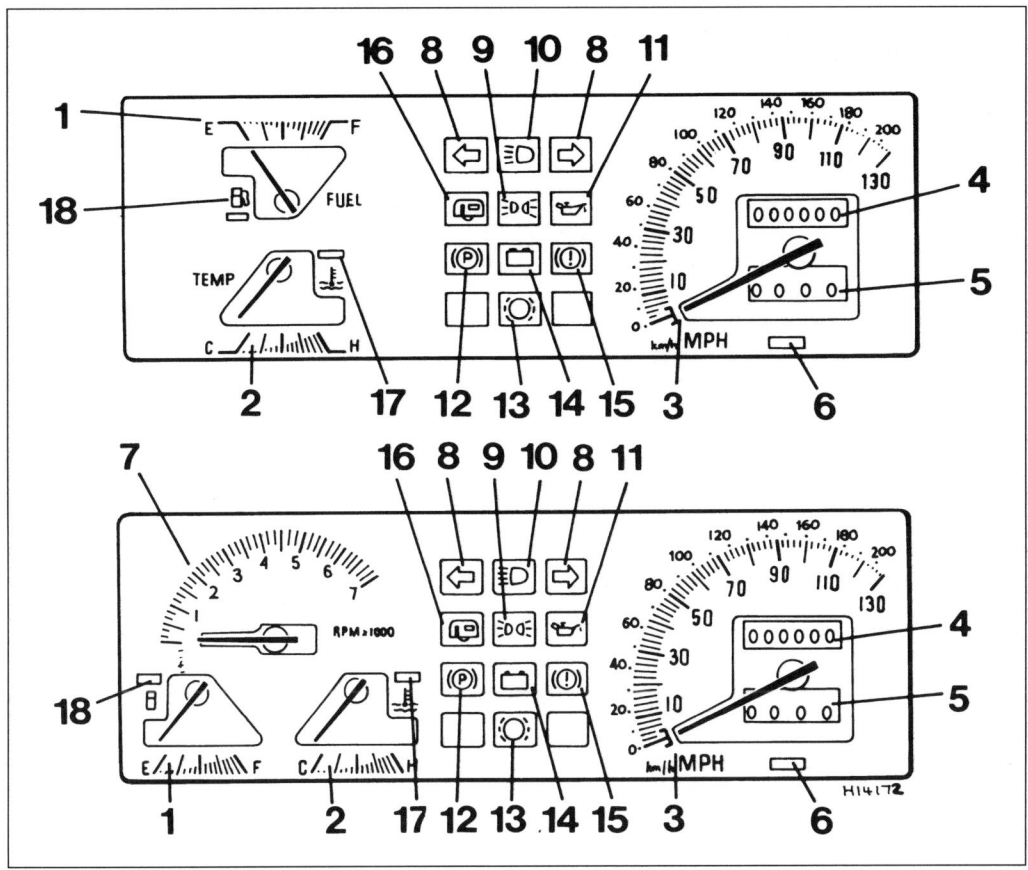

▲ Instrument panel layout – later Maestro, all other Montego models

1	Fuel gauge	10	Main beam 'ON' warning lamp
2	Temperature gauge	11	Low oil pressure warning lamp
3	Speedometer	12	Handbrake 'ON' warning lamp
4	Total mileage recorder	13	Brake pad wear warning lamp
5	Trip mileage meter	14	Ignition/no-charge warning lamp
6	Trip meter reset button	15	Low brake fluid level warning lamp
7	Tachometer ('rev. counter')	16	Trailer direction indicator warning lamp
8	Direction indicator/hazard warning lamps	17	High coolant temperature warning lamp
9	Sidelamps 'ON' warning lamp	18	Low fuel level warning lamp

DRIVER'S INSTRUMENTS AND CONTROLS

Speedometer

Indicates the car's road speed, either as a digital display or by means of a conventional analogue dial, and incorporates a total mileage recorder. All models also have a trip computer (see below) or a trip mileage meter, which can be reset to zero by pressing the button protruding from the instrument.

Tachometer ('rev. counter')

Indicates engine speed in revolutions per minute (rpm), either by the progressive illumination of coloured segments on a scale, or by means of a conventional analogue dial. For normal driving and the best fuel economy, the engine speed should be kept in the 2000 to 4000 rpm range. The red sector indicates the maximum permissible engine speed; the needle should NOT be allowed to enter this sector under any circumstances.

Fuel gauge

Indicates the quantity of fuel remaining in the tank.

Where a conventional gauge is used, the empty end of the scale is marked 'E' and is shown by the red (or yellow) sector, while the full end of the scale is marked 'F'; on some models a warning lamp, which will light if the level falls very low, is mounted next to the gauge. When the needle approaches the empty end of the scale, the fuel level is low and the tank should be refilled as soon as possible; if the warning lamp should light, fill up immediately.

Where the fuel level is shown by a scale of coloured segments, all will be lit when the tank is full, and will progressively go out as fuel is used; when only the red segments remain lit, the fuel level is very low, and additional warnings will be displayed or given.

Temperature gauge

Indicates the engine coolant temperature.

Where a conventional gauge is used, on early models the cold end of the scale is shown by the blue sector, the hot end by the red sector. On later models, the cold end of the

scale is marked 'C' while the hot end is marked 'H' and has yellow graduations; on some models a red warning lamp, which will light if the temperature rises to an excessive level, is mounted next to the gauge.

As the engine warms up, the needle should move from the cold end of the scale into the middle region. If the needle remains in the cold end, or enters the hot end (especially if the warning lamp lights), a fault is indicated and advice should be sought (refer to *'Fault finding'* on page 139).

Where engine temperature is shown by a scale of coloured segments, normal running temperature is shown when all segments are lit to the middle of the scale; if the temperature should rise higher, red segments will light and additional warnings will be displayed or given.

Do not continue to run an engine which shows signs of overheating.

Clock

The clock starts automatically, and must be reset whenever the battery is disconnected.

If the display cannot be seen clearly enough, switch on the ignition while adjustments are made.

Using a pointed instrument such as a ballpoint pen, press the 'H' or 'HRS' button to change the hour setting, and press the 'M' or 'MINS' button to change the minutes setting.

▲ *Setting the clock*

To zero the display, press both buttons simultaneously.

Where a 24-hour mode clock is fitted, to transfer from one mode to the other, zero the display – if the clock is in 12-hour mode, it will display '01.00', if in 24-hour mode the display will be '00.00'. If the clock is in the wrong mode, zero the display again – the time can then be set as described. If the battery is disconnected, the clock will revert to 12-hour mode when it restarts.

Trip computer and voice synthesis unit – Maestro

Provides information on current time and date, journey distance and fuel consumption. Once the system has been reset at the start of a journey by pressing the 'Clear' button, information is given (whenever one of the other function buttons is pressed) by a display which can be either in imperial or in metric units of measurement, depending on which has been selected for the speedometer; otherwise the display will show the current time.

A voice synthesis unit provides a range of spoken messages and warnings, each preceded by a two-tone chime and interrupting any in-car audio equipment in use, to reinforce the information displayed; the unit can be switched on or off, and the sound volume can be adjusted by the 'On-Vol' control. If the 'Restrict' button is pressed, the voice unit will issue warnings only; press again to restore full operation.

The function buttons are as follows – whenever a button is pressed, the information required will be displayed, the voice unit will broadcast a message confirming the function selected, and the red indicator lamp above the button will light:

TIME – When pressed once, displays the current time in hours and minutes. When pressed twice (once only, if 'Time' is already selected), displays (for three seconds only) the current date in days and months. The clock will function, and can be reset, as described previously; if the date is to be reset, press the 'Time' button twice, then depress the clock's 'H' button until the day is correct, and the 'M' button until the month is correct – work quickly, as the display will revert to the current time after three seconds.

DIST – Displays the distance travelled on a journey (ie, since the 'Clear' button was last pressed).

CLEAR – Clears display and resets system. Must be pressed for at least three seconds to operate (a safety factor is built in to the system, to avoid information being cleared by momentary accidental pressing of the button).

FUEL – Displays the fuel used for a journey (ie, since the 'Clear' button was last pressed).

INST – Displays (in miles/gallon or litres/100 km) the actual fuel consumption at the moment the button is pressed. The reading given depends on road conditions and throttle opening, and so may not be the same as the AVE reading.

AVE – Displays the average fuel consumption for a journey (ie, since the 'Clear' button was last pressed). If the unit's capacity is exceeded, the display will show 'OFLO' (overflow) until the 'Clear' button is depressed for at least three seconds.

Press the 'Fuel/Inst/Ave' buttons simultaneously to check the full range of voice messages – all messages will be broadcasted progressively (unless the 'Restrict' button is depressed, in which case only the warnings will be given) – press 'Time/Ave' simultaneously to test the voice messages (as above) and instruments (half-value readings will be displayed).

Driver Information System – Montego

The Driver Information Centre is a more sophisticated version of the Maestro's trip computer and voice synthesis unit; it also incorporates all the usual instrument functions in an electronic display, and has a service interval indicator to remind the driver of when the next service is due.

A vehicle map shows which lamps are switched on, and warns of open doors and/or boot lid; if a lamp bulb fails, the appropriate section of the map will go out.

In addition to the spoken warnings, the message centre will display the appropriate warning and the 'Attention display' will flash to reinforce the point.

A comprehensive instruction booklet was supplied with the car when new, and is essential for getting the best out of the system; if the booklet is no longer available, a Rover dealer may be able to order a copy for you.

The basic operation of the various controls is outlined below:

CLOCK/ODOMETER

Press the 'Select' button until time and distance are displayed. To reset the clock, press 'Reset' once to convert the display to 24-hour mode – a dash will show next to the hour figure. Press the 'Set' button until the hour figure is correct, then press 'Reset' to move the dash to the minutes figure; press 'Set' until the minutes figure is correct, then press 'Reset' again to return the display to 12-hour mode. The colon between the hour and minutes figures will flash to show that the clock is operating.

CALENDAR

The calendar displays date, month and year up to the year 2079, but will not recognise leap years. Press the 'Select' button to display the calendar. To change the date, follow the same procedure used to reset the clock, progressing from day to month to year; when the calendar is correctly reset, press 'Reset' a fourth time.

DISPLAY UNITS

The display can be either in imperial or in metric units of measurement; press the 'Imp/Met' button to switch from one to the other.

RESTRICTION OF FUNCTIONS

Press the 'Restrict' button briefly to limit the visual display to the speedometer and the lowest (red) segment of the fuel gauge. Press the button for three seconds to restrict the display and the voice – spoken warnings only will then be issued; press the button again (briefly) to restore the display (for three seconds) to restore full operation.

TRIP COMPUTER

The Montego unit has the same four basic functions as the Maestro version previously described. Press the 'Select' button and keep it depressed to move the display through the clock, calendar and all trip computer functions at half-second intervals; pressing the 'Inst' and 'Ave' buttons will give the appropriate message without delay.

To zero all computer functions, select any function and press the 'Reset' button for at least three seconds.

SERVICE INTERVAL INDICATOR

To reset the service interval indicator, first check that the calendar is correctly set (see above).

Switch off the ignition, hold the 'Select' button depressed and switch on again; the message centre should display 'ENGLISH' – press 'Select' again and the display will clear – press 'Reset' and hold it depressed until the display shows 'OIL CHANGE' – release the 'Reset' button, then press it again and hold it depressed until the display shows 'SERVICE RESET', then release the 'Reset' button – press 'Select' to finish.

Econometer

This is a vacuum gauge which, when switched on, uses coloured Light-Emitting Diodes (LEDs) to indicate whether or not the engine is within its most economical operating range. If the centre (green) LED only is lit, then fuel economy will be at its best. If yellow shows, especially when accelerating, economy will be reasonable, but if red shows you should modify your driving style (either by releasing slightly the accelerator pedal or by changing gear, as appropriate) for better economy.

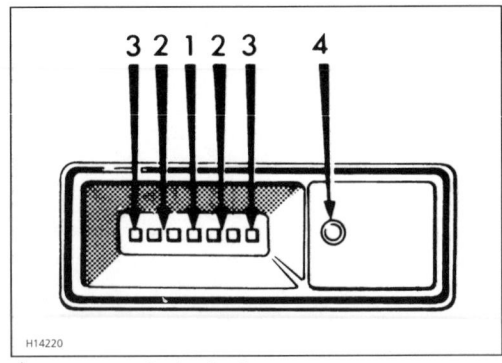

▲ Econometer display details
1　Green – good
2　Yellow – reasonable
3　Red – poor
4　On/off button

Warning lamps

These lamps warn the driver of a fault, or inform the driver that a particular device is in operation. When the ignition is first switched on, the ignition/no-charge and low oil pressure warning lamps will light, then will go out on starting the engine; note that all braking system warning lamps will light whenever the handbrake is applied. The function of each individual lamp is as follows.

HANDBRAKE 'ON' WARNING LAMP

Warns that the handbrake is applied. If the lamp lights while driving, check that the handbrake is fully released.

LOW BRAKE FLUID LEVEL WARNING LAMP

Warns that the brake fluid level is low. If the lamp lights whilst driving, stop immediately and check the brake fluid level (refer to 'Regular checks' on page 84). Note that the circuit can be checked by (with the ignition switched on and the handbrake released) depressing the flexible contact cover in the centre of the brake fluid reservoir filler cap; the lamp should light. If it does not light, seek advice; either the bulb is blown, or there is a fault in the circuit.
   Do not continue to drive the car if a brake fluid leak is suspected.

LOW OIL PRESSURE WARNING LAMP

Warns that the engine oil pressure is low. If the lamp stays on for more than a few seconds after start-up, it's likely that the engine is worn, and advice should be sought. If the lamp lights whilst driving, switch off the engine immediately and seek advice.

IGNITION/NO-CHARGE WARNING LAMP

Acts as a reminder that the ignition is switched on if the engine isn't running, and also serves as a no-charge warning lamp. If the lamp stays on after starting, or lights whilst driving, the battery is not being properly charged. The battery may therefore become fully discharged (resulting in a 'flat' battery), so the best course of action is to stop and seek advice.

CHOKE WARNING LAMP

Acts as a reminder that the choke is in operation. The lamp will go out when the choke control is pushed fully in.

MAIN BEAM 'ON' WARNING LAMP

Acts as a reminder that the headlamp main beam is switched on.

SIDELAMPS 'ON' WARNING LAMP

Acts as a reminder that the sidelamps and tail lamps are switched on.

DIRECTION INDICATOR WARNING LAMP(S)

Show(s) that the direction indicators are switched on.

BRAKE PAD WEAR WARNING LAMP

Indicates that the front brake pads are worn to the stage where they require renewal. If the lamp lights, the pads should be renewed as soon as possible.

HIGH COOLANT TEMPERATURE WARNING LAMP

Reinforces the temperature gauge's message.

LOW FUEL LEVEL WARNING LAMP

Reinforces the fuel gauge's message.

SEAT BELT WARNING LAMP

Lights if the ignition is switched on and either the driver's or (front) passenger's seat is occupied without the belt(s) being fastened.

TRAILER DIRECTION INDICATOR WARNING LAMP

If the car has a towbar installed, so that a heavy-duty trailer flasher unit (relay) has been fitted, this lamp shows that the direction indicators are switched on. If trailer lighting is connected at the towbar socket, the lamp will flash in unison with the direction indicator warning lamp; if trailer lighting is not connected, the lamp will flash once only each time the direction indicators are switched on, to show that the circuit is sound.

Ignition switch/steering lock

The switch has four positions as follows:
- **O** Ignition off, steering locked
- **I** Ignition off, accessory circuits on, steering unlocked
- **II** Ignition on, all electrical circuits on
- **III** Starter motor operates (release the key immediately the engine starts)

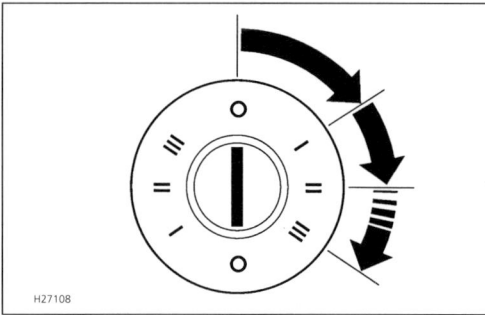

▲ *Ignition switch/steering lock key positions*

Fuel cut-off inertia switch – fuel injection (EFi) models

To reduce the risk of fire after an accident, an inertia-operated switch is fitted to cut off the electrical supply to the fuel pump of all fuel-injected models, in the event of any sudden and sharp deceleration. If the engine stops and fails to restart after the car has been braked sharply enough to throw the occupants forward against their seat belts, the switch may well have tripped.

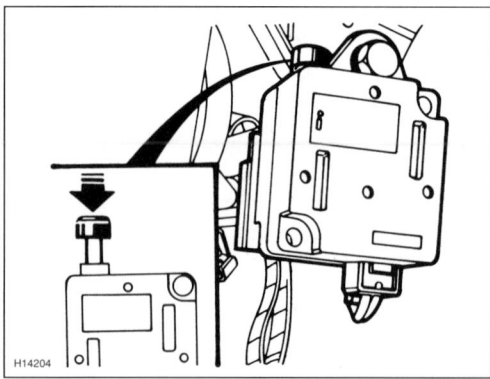

▲ *Fuel cut-off inertia switch – fuel-injected models*

The switch can be found behind the right-hand side of the centre console, forwards of the radio/cassette player on Maestro and early Montego models. On later Montego models, remove the fusebox cover (refer to *'Fuses'* in *'Bulbs, fuses and relay renewal'* on page 127); the switch is next to the fusebox panel.

If the switch has tripped, a red stem will be visible projecting from the switch body. To reset the switch, first check for safety's sake that there are no damaged fuel lines (which should be evident from the smell of petrol and/or traces of leakage on the ground under the car), then depress the button. The engine should then start.

Choke

The choke control should be pulled out as necessary to start a cold engine, and should be gradually pushed in as the engine warms up, until the car can be driven smoothly with no choke. The amber warning lamp will light when the choke is in use.

Gearbox

MANUAL GEARBOX

The gear positions follow the usual 'H' pattern. Never rest your hand on the gear lever while driving, or you will cause premature wear of the selector mechanism.

Reverse must only be engaged when the car is stationary, with the clutch pedal fully depressed – pause, to eliminate crunching, before moving the gear lever. On 2.0 litre models and 1988-on 1.6 litre Montego models, reverse gear is selected by moving the lever fully to the right against spring pressure and pulling it rearwards. On all other models, reverse is selected by moving the lever fully to the left against spring pressure and depressing it until it can be pushed further to the left and forwards. Where applicable, when changing down from 5th gear, do not press the lever to the left, or 2nd gear will be selected instead of 4th. On 1.3 & 1.6 litre models with a five-speed gearbox, 5th gear selection pressure is adjustable; if you experience difficulty selecting this gear, seek the advice of a Rover dealer.

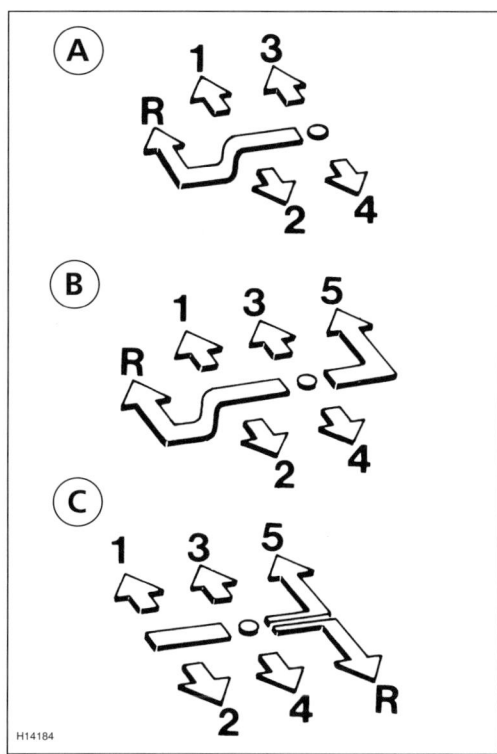

▲ Gear lever positions (manual gearbox)

A Four-speed gearbox

B Five-speed (Volkswagen) gearbox –
1.3, Maestro 1.6 and early Montego 1.6 models

C Five-speed (Honda/Rover) gearbox –
all 2.0 and later Montego 1.6 models

AUTOMATIC TRANSMISSION

1.6 litre models

There are 6 selector positions:

P – locks the transmission mechanically, and should be used when parking. The engine can be started, but must not be run above fast idle speed. Only select **P** when the car is stationary and the handbrake has been applied; depress the button on the side of the selector handle to select or move out of this position.

R – selects reverse gear. Only select **R** when the car is stationary; depress the button on the side of the selector handle to select this position.

N – selects neutral. The engine can be started, but must not be run above fast idle speed. No power is transmitted to the wheels.

D – selects **Drive**. The transmission will change gear automatically using all three gears, with down-changes for increased acceleration possible, when required, through a kickdown linkage operated by pressing the accelerator pedal fully to the floor. **D** can be selected either when the car is stationary or when it is moving. In normal use **D** will, with the kickdown facility, give full control of the car, with no need for the driver to worry about gear-changing, but drivers should remember that there is no engine braking in the first ratio when **D** is selected. The transmission does, however, permit manual override when required; starting off in position **1**, select **2** and then **D** as road speed increases (remembering the maximum change speeds). When changing down, remember the lack of engine braking in position **1**.

2 – locks the transmission in 1st and 2nd gear; has engine braking only in 2nd. To be selected when the car is being driven, but **never** at speeds of more than 70 mph (110 km/h). Useful when driving up or down hills, or for towing.

1 – locks the transmission in 1st gear, and has engine braking; depress the button on the side of the selector handle to select this position. In normal use, should be selected only when the car is stationary; if selected while driving, **never** do so at speeds of more than 40 mph (65 km/h). It is useful when driving up or down extremely steep slopes, or when towing at low speeds.

The handbrake or footbrake must be applied before selecting any gear when the car is

▲ Automatic transmission selector positions –
1.6 litre models

Note: Press button (arrowed) to select positions indicated

stationary; do not run the engine above normal idle (tickover) speed when the car is stationary and a gear is selected. For very short halts, the transmission can be left in gear, provided that the engine remains at idle speed and the brakes are applied; for longer halts, position **N** should be selected to save wear and minimise fuel consumption.

2.0 litre models

These models are fitted with a transmission which has three normal forward speeds and an overdrive. Apart from the differences noted below, it is operated as described above for the three-speed unit.

There are 7 selector positions:

D4 – selects **Drive**. The transmission will change gear automatically using all three gears and the overdrive. Use this position for open road and motorway driving, but remember that there is no engine braking in the first ratio when **D4** is selected.

D3 – locks the transmission in 1st, 2nd and 3rd gears. Use this position for normal driving.

2 – locks the transmission in 1st and 2nd gear; has engine braking only in 2nd. To be selected when the car is being driven, but **never** at speeds of more than 55 mph (90 km/h).

1 – locks the transmission in 1st gear, and has engine braking; depress the button on the side of the selector handle to select this position. In normal use, should be selected only when the car is stationary; if selected while driving, **never** do so at speeds of more than 30 mph (50 km/h).

Lighting switch – early Maestro models

Move the switch upwards, towards the steering wheel, to the first position to switch on the sidelamps (the instrument panel lamps and switch illumination should also light), and to the second position to switch on the headlamps.

Instrument illumination rheostat

Rotate the control knob to alter the intensity of the instrument display and illumination.

Headlamp dip/flash, horn and direction indicator switch – early Maestro models

Controls the headlamp dip/main beam and flash, horn and direction indicators.

Press the dipswitch towards the facia for main beam; lift it towards the steering wheel to flash the headlamp main beams. Move the stalk up or down, as desired, to indicate a change of direction; press in its tip to sound the horn.

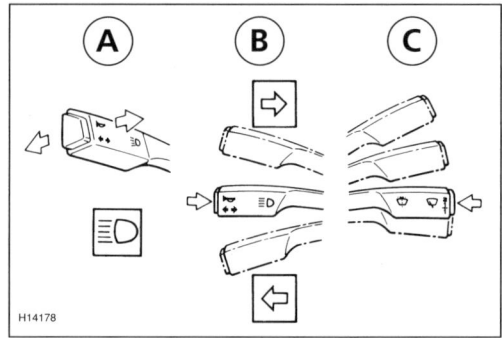

▲ *Multi-function switches – early Maestro models*
  A *Dipswitch/headlamp flash*
  B *Direction indicators/horn*
  C *Windscreen wiper/washer*

Lighting, headlamp dip/flash, horn and direction indicator switch – all except early Maestro models

Controls the direction indicators, horn, sidelamps and headlamps, headlamp flash and dip/main beam.

On later models, if the ignition and sidelamps are switched on together, the headlamp dip beams will light at one-sixth of their normal output to prevent the car from being driven on sidelamps alone ('dim-dip' lighting). Also, on later Montego models, a warning buzzer will sound if the lights are left on when the ignition is switched off and one of the doors is opened.

To indicate a lane change (on motorways etc) just press the switch lightly in the direction desired, and hold it against spring pressure until the manoeuvre is completed.

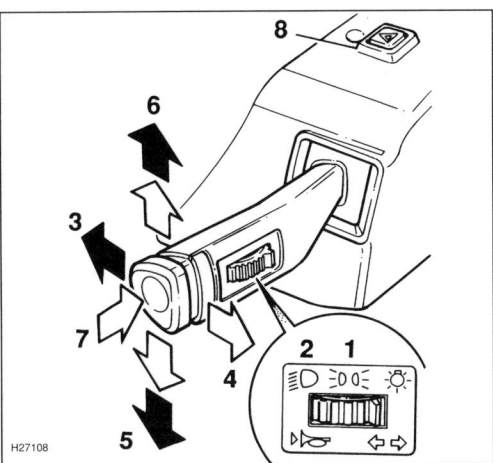

▲ Lighting, headlamp dip/flash, horn and direction
indicator switch and hazard warning switch –
all except early Maestro models
1 Lighting switch sidelamp position
2 Lighting switch headlamp position
3 Main beam (towards facia)
4 Headlamp flash (towards steering wheel)
5 To indicate left turns/lane change
6 To indicate right turns/lane change
7 Horn
8 Hazard warning

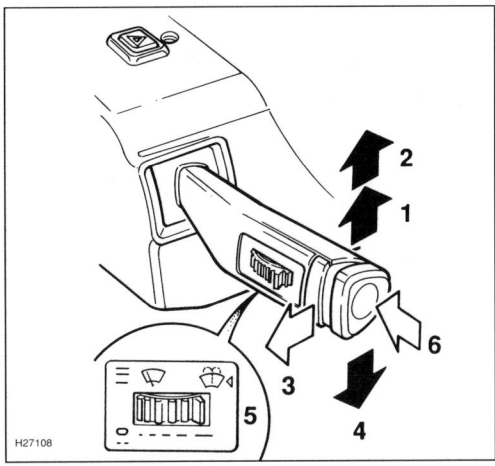

▲ Windscreen wash/wipe switch –
all except early Maestro models
1 Normal wiper speed
2 Fast wipe
3 Single 'flick' wipe
4 Intermittent wipe (where fitted)
5 Intermittent wipe delay control (where fitted)
6 Windscreen washers (and, where fitted,
programmed wash/wipe)

Windscreen wash/wipe switch – early Maestro models

Controls the windscreen wipers and washers.
Press down for single 'flick' wipe or (where
fitted) intermittent wipe, press up once for
normal wiper speed, to the second position
for fast wipe. Press in the tip to operate
the washers.

Note: *Refer to the illustration on page 26.*

Windscreen wash/wipe switch – all except early Maestro models

Controls the windscreen wipers and washers.
On models with intermittent wipe fitted, vary
the delay between wipes by rotating the
control **(5)** to the right to increase the delay, to
the left to reduce the delay. If a programmed
wash/wipe system is fitted, pressing the switch
(6) will override any other wiper function, and
operate the washers and wipers for as long as
the switch is depressed; on releasing the
switch, the washers will cease immediately
while the wipers continue for five seconds.
Where headlamp washers are fitted, these will
operate automatically when the windscreen
washers are used, if the headlamps are
switched on.

Tailgate wiper switch

Press in the switch to operate the wiper. The
switch warning lamp (where fitted) will light,
and the wiper will operate continuously,
unless the intermittent type is fitted; these
operate continuously for ten seconds, then
once every six seconds. Press the switch again
to stop the wiper.

Tailgate washer switch

Press in the switch to operate the washer,
press it again to stop it; the switch warning
lamp (where fitted) will light while the washer
is operating.

Heater/air distribution controls

EARLY MAESTRO MODELS

Control **1** – controls the supply of fresh, unheated, air to the face level ventilation vents: closed off at the top position.

Control **2** – controls the temperature of air from the heater: up (blue) for cold, down (red) for hot, with variable settings in between.

Control **3** – controls the distribution of air from the heater.

> Top position – Air directed to the windscreen and (where fitted) side window demist vents **(7)**.
> Bottom position – Air directed mainly to the footwell vents, less to the windscreen vents.

Control **4** – controls the air supply by operating the blower fan.

> Position 0 – Air intake to heater closed off.
> 2nd position – Blower off, air supply by the ram effect of the car moving forwards.
> 3rd position – Blower at slow speed.
> Position 2 – Blower at normal speed for normal heating and ventilation.
> Position 3 – Blower at high speed for maximum heating and ventilation.

Controls **5** – adjust the direction of air from the face level ventilation vents.

Controls **6** – control the volume of air from the face level ventilation vents.

LATER MAESTRO MODELS

Control **1** – controls the temperature of air from the heater: up (blue) for cold, down (red) for hot, with variable settings in between.

Control **2** – controls the distribution of air from the heater; the control may be set between positions as required.

> Top position – Fresh air directed to face level vents only.
> 2nd position – Fresh air directed to the face level vents, reduced airflow (which can be heated) to the windscreen vents.
> Centre position – Air directed to the windscreen and (where fitted) side window demist vents.
> 4th position – Air directed to the windscreen and footwell vents.
> Bottom position – Air directed mainly to the footwell vents, less to the windscreen vents.

Switch **3** – controls the air supply by operating the blower fan.

> Position '0' – Blower off, air supply by the ram effect of the car moving forwards.
> Position 'I' – Blower at slow speed.
> Position 'II' – Blower at normal speed for normal heating and ventilation.
> Position 'III' – Blower at high speed for maximum heating and ventilation.

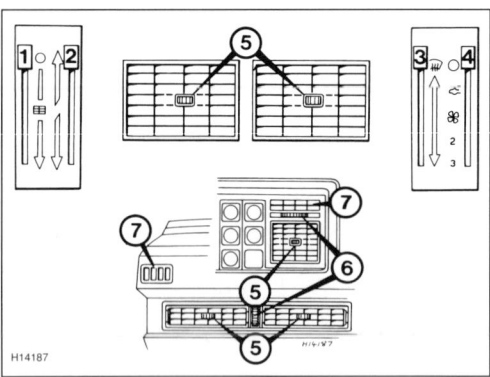

▲ *Heater/air distribution controls – early Maestro models*
 1 Face level ventilation control
 2 Temperature control
 3 Air distribution control
 4 Air supply control
 5 Vent adjusting controls
 6 Vent air volume controls
 7 Side window demist vents

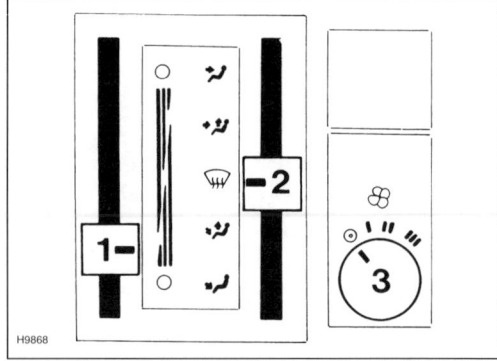

▲ *Heater/air distribution controls – later Maestro models*
 1 Temperature control
 2 Air distribution control
 3 Air supply (blower) switch

EARLY MONTEGO MODELS

Control **1** – controls the air supply.
  Top position (O) – Air intake to heater closed off.
  1st position – Blower off, air supply by the ram effect of the car moving forwards.
  2nd position – Blower at slow speed.
  3rd position – Blower at normal speed for normal heating and ventilation.
  4th position – Blower at high speed for maximum heating and ventilation.
  Control **2** – controls the temperature of air from the heater: up (blue) for cold, down (red) for hot, with variable settings in between.
  Switch **3** – controls the distribution of air. Normal air distribution balances the flow of air between the windscreen and footwell vents. To shut off the footwell vents, directing all the flow to the windscreen for demisting/defrosting, press in the switch; its amber warning lamp will light. Press the switch again to restore normal ventilation.

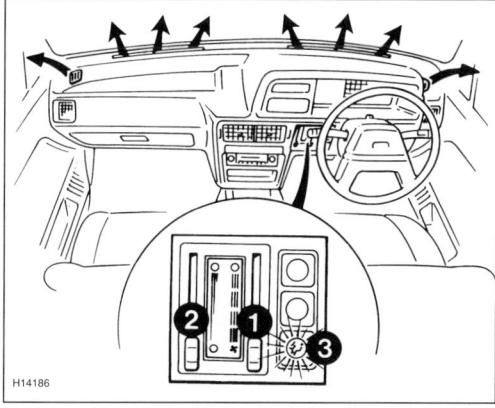

▲ Heater/air distribution controls – early Montego models
  1 Air supply (blower) control
  2 Temperature control
  3 Air distribution switch

LATER MONTEGO MODELS

Left-hand knob – controls the distribution of air from the heater; the control may be set between positions as required.

Rotated fully to the left – Air directed mainly to the footwell vents, less to the windscreen vents.
  Centre position – Air directed to the windscreen and side window demist vents.
  Rotated fully to the right – Fresh air directed to face level vents only.
  Centre switch – controls the air supply by operating the blower fan.
  1st position – Blower off, air supply by the ram effect of the car moving forwards.
  Position 'I' – Blower at slow speed.
  Position 'II' – Blower at normal speed for normal heating and ventilation.
  Position 'III' – Blower at high speed for maximum heating and ventilation.
  Right-hand knob – controls the temperature of air from the heater: rotated fully to the left (blue) for cold, to the right (red) for hot, with variable settings in between.

Hazard warning switch

Operates all the direction indicator lamps simultaneously, regardless of the position of the ignition switch. Press the switch down to switch on; the switch warning lamp will flash in unison with the direction indicator warning lamp(s). Press the switch again to switch off.

Rear foglamp switch

Operates the rear foglamp(s), provided that the headlamps are switched on. Press the switch in to switch on; the switch warning lamp will light. Press the switch again to switch off. Use the rear foglamps **only** when visibility is seriously reduced, to 100 yards or less, by foggy conditions; **do not** use them simply because it is dark, raining or misty.

Heated rear window switch

Operates the heated rear window, provided that the ignition is switched on. Press the switch in to operate the heater; the switch warning lamp will light. Press the switch again to stop the heater. It's recommended that the heater is switched off as soon as demisting is complete, as drain on the battery is very high; note that on later models, the heater is cut out automatically after approximately ten minutes.

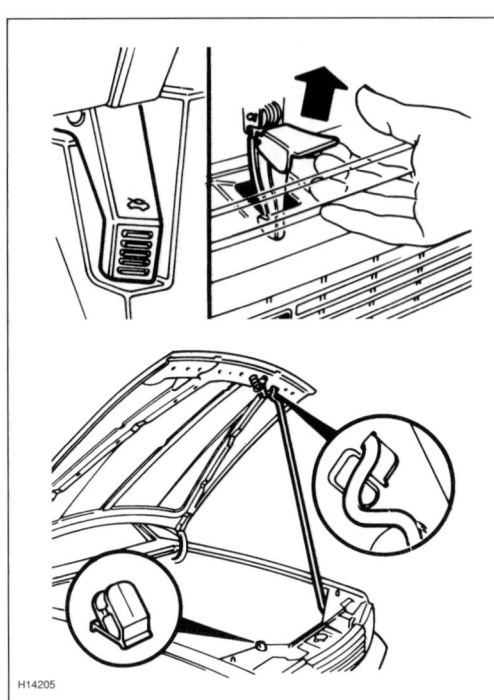

H14205

▲ *Bonnet release mechanism – typical*

H14188

▲ *Remote boot lid release*
 1 Release lever
 2 Remote release inhibitor lever –
   operational (white), disabled (black)

Bonnet release lever

Mounted under the facia right-hand end. Pull the lever to release the bonnet, then lift the bonnet safety catch (slightly to the right of centre) to open the bonnet.
 When closing the bonnet, ensure that the safety catch is engaged.

Boot lid release lever

Lift the lever **(1)** located at the side of the driver's seat to open the boot lid from within the car. To disable this, press forwards the lever **(2)** beneath the boot lid lock; the boot can now be opened only by key.

Fuel filler flap release lever

Lift the lever located at the side of the driver's seat to open the filler flap from within the car.

INTERIOR EQUIPMENT

Cigar lighter

Operates regardless of the position of the ignition switch. Press the lighter in, then release it, and wait until it pops out ready for use.

Electric windows

MAESTRO

On early models, when the ignition is switched on, switches mounted in the front door armrests raise and lower the front windows. Two switches are provided on the driver's side, so that the driver can control the passenger's side window. Press the switch forward to raise the window, to the rear to lower it; keep the switch pressed until the window reaches the desired position.
 On later models, the switches are mounted in the centre console (the left-hand switch controlling the left-hand window). Press a switch down and release to lower fully the window. To raise the window, press the switch up and hold until the window reaches the required position; to close the window, press the switch up again and hold it until the window is fully closed.

MONTEGO

On early models where only the front windows are electrically-operated, switches mounted in the front door armrests raise and lower the front windows; two switches are provided on the driver's side, so that the driver can control the passenger's side window.

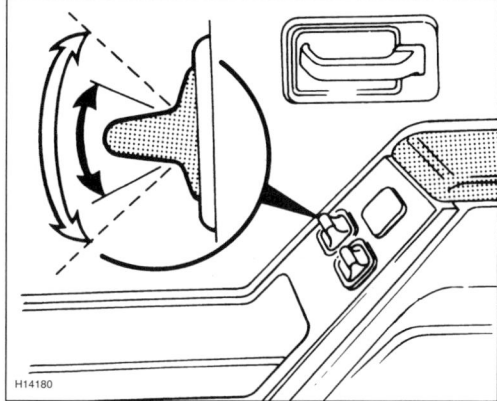

▲ Early Montego model electric window switches – front only. **Note:** Inset shows 'one-touch' operation for driver's window

Where all four windows are electrically-operated, five switches are mounted on the centre console, with a further switch in each rear door's armrest; an isolation switch **(5)** can be pressed up to prevent operation of the rear windows; press the switch down to restore operation. Press a switch up to raise the window, down to lower it; either keep the switch pressed lightly until the window reaches the desired position, or (driver's door window only) move the switch to its second position, and release to open or close the window fully.

On later models, the switches are mounted in the centre console (a left-hand switch controlling the appropriate left-hand window), with the rear window isolation switch mounted in the facia, on the right of the steering column; operation of all windows is possible up to 45 seconds after the ignition has been switched off. To open one of the front windows fully, press and release the rear of the appropriate switch; to raise the window, press the switch up and hold until the window reaches the required position; to close the

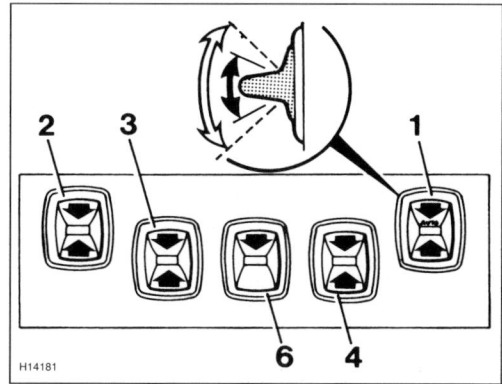

▲ Early Montego model electric window switches – front and rear windows
 1 Driver's door window switch – inset shows 'one-touch' operation
 2 Front passenger door window switch
 3 Rear left-hand passenger door window switch
 4 Rear right-hand passenger door window switch
 5 Rear window isolation switch

window, press the switch up again and hold it until the window is fully closed. To open one of the rear windows, press the rear of the switch, and press the front of the switch to raise it; keep the switch pressed until the window reaches the desired position.

Sunroof
MANUALLY-OPERATED

To open the sunroof on early models, pull the handle **(1)** down and depress pushbutton **(2)**

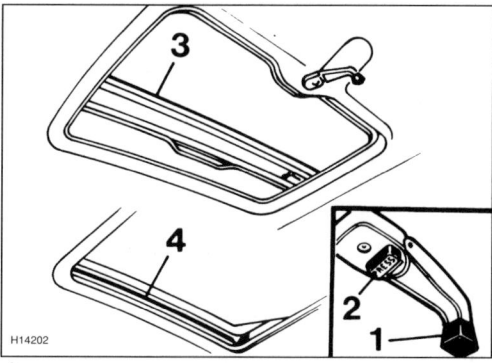

▲ Manually-operated sunroof – early models
 1 Handle   **3** Roof panel open
 2 Pushbutton   **4** Roof panel rear edge raised

to lock in the operating position; on later models, pull the handle from its recess and swing it rearwards, then upwards until it locks.

Rotate the handle anti-clockwise to open the roof panel (and sun visor, where fitted), clockwise (until a click is heard, on later models) to close it; where a sun visor is fitted, this must be pulled forwards manually to close it.

To tilt the sunroof's rear edge upwards (from the closed position) for ventilation, unfold the handle as described above and rotate it clockwise; turn anti-clockwise (until a click is heard, on later models) to lower it again.

ELECTRICALLY-OPERATED

The sunroof can be operated when the ignition is switched on, and for up to 45 seconds after the ignition has been switched off. Open the roof by pressing the rear of the switch, close by pressing the front of the switch; keep the switch pressed until the roof reaches the desired position. The sun visor must be pulled forwards manually to close it.

To tilt the sunroof's rear edge upwards (from the closed position) for ventilation, press the front of the switch; press the rear of the switch to lower it again.

To close the roof panel in the event of electrical failure, press forwards the rear edge of the panel next to the switch, and swing the panel down to expose a large nut. Using the 21 mm box spanner and emergency key (supplied in a pouch stowed in the glovebox with the car when new), slacken the nut (anti-clockwise) through three full turns, then insert the key into the hexagonal hole in the centre of the nut, and rotate the key anti-clockwise to close the roof, or clockwise to lower its rear edge. Have the circuit repaired and the nut retightened by a Rover dealer.

Note: *The tightness of the nut is critical to the correct operation of the sunroof – do not attempt to adjust it yourself, or you may damage the opening mechanism.*

Interior rear view mirror

To prevent the mirror from becoming detached from its adhesive pad, always hold the base firmly before adjusting the mirror position.

If a dipping mirror is fitted, pull back the lever under the mirror head to reduce the glare from the lights of following vehicles at night.

Exterior rear view mirror(s)

The mirror can be adjusted by swivelling its head on the base; on some models, a remote control is provided so that adjustment can be made from inside the car, either manually or electrically, without opening the window.

Note: *Mirrors with electrically-operated adjustment also have heating elements built into the glasses that operate whenever the ignition is switched on.*

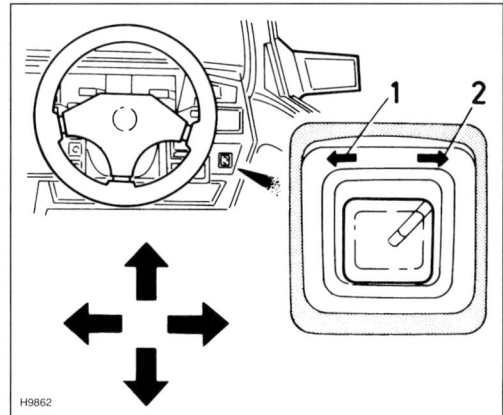

▲ *Electrically-operated exterior mirror adjustment*
  1 Passenger's door mirror selector position
  2 Driver's door mirror selector position

Seat belts

The operation of the seat belts is self-explanatory for most models, but certain points require additional explanation.

FRONT SEAT BELT ADJUSTMENT

On early Maestro models, a limited degree of adjustment is possible for the front seat belt upper anchorage; see a Rover dealer for details.

On later Maestro and all Montego models, pull out the knob **(5)** and move the anchorage until the belt passes over the wearer's shoulder, mid-way between the neck and the edge of the shoulder; release the button and check that the anchorage is locked.

REAR SEAT BELTS – ALL MODELS

The centre rear seat belt is of the lap type, and must be manually adjusted for fit by the wearer. To tighten the belt, pull the webbing through the adjuster, and move the plastic clip to hold the end of the belt.

If the rear seats are to be folded forwards, be patient and disengage the seat belts from the seats before moving them, so that there is no risk of damage either to the seats or to the belts; when replacing the seats, ensure that the belts are routed correctly between the seat base and back cushions, and that the webbing is not trapped or damaged (see below).

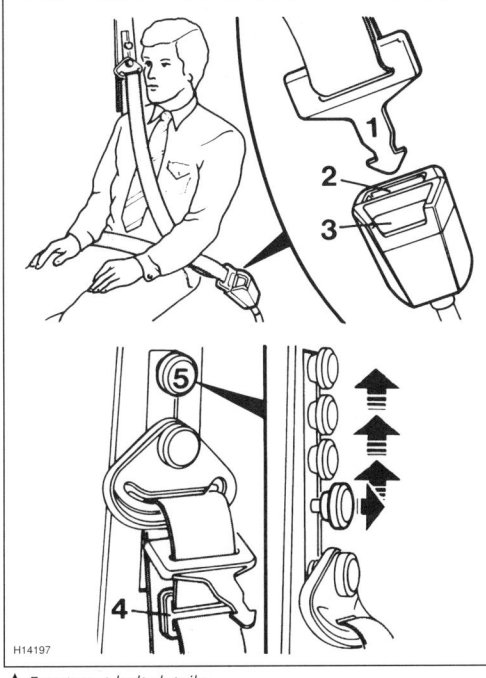

▲ Front seat belt details
1 Tongue   4 Slider
2 Lock   5 Upper anchorage release knob
3 Release

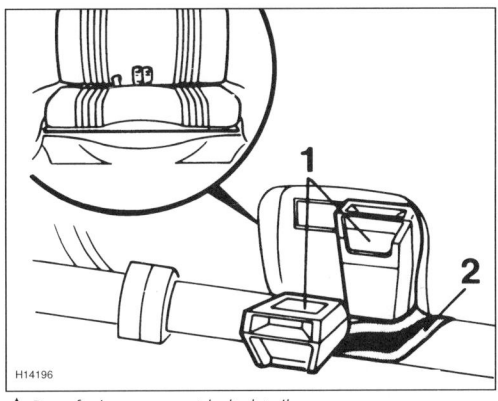

▲ Rear-facing rear seat belt details
1 Locks   2 Elastic strap

Front seats

FRONT/REAR MOVEMENT

Pull up the bar **(1)** located under the front of the seat, then slide the seat to the required

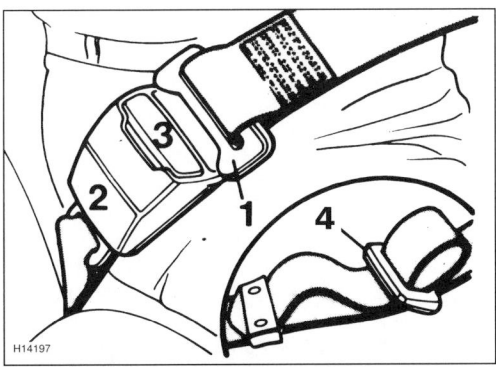

▲ Rear centre seat belt details
1 Tongue   3 Release
2 Lock   4 Adjuster

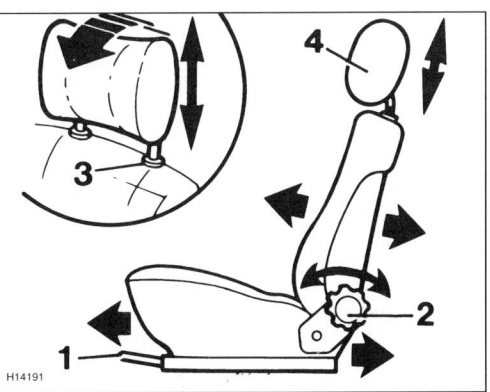

▲ Front seat adjustment
1 Seat release bar
2 Seat back adjuster wheel
3 Head restraint release – where fitted
4 Head restraint adjustment

position. Release the bar, and ensure that the seat has locked in position by gently rocking it backwards and forwards.

SEAT BACK ADJUSTMENT

Taking the weight of your body off the seat back, turn the wheel **(2)** on the inside bottom edge of the seat to alter the angle of the seat back.

LUMBAR SUPPORT

Rotate the handwheel on the side of the seat back to adjust the lumbar support as required.

HEAD RESTRAINTS

The head restraints can be adjusted for height by moving them up or down in their guides; on later models with the full-cushion type of head restraint, unlock first each fitting by depressing the release **(3)**. Position the restraint so that it supports the back of your head, **not** your neck.

Rear seats

MAESTRO

Either a full-width or a split rear seat may be fitted, depending on model.

To fold the seat back forwards, open the tailgate and release the locking levers **(1)** at each upper end of the seat, under the parcel shelf supports. Unhook the parcel shelf support straps from the tailgate, fold the seat back forwards on to the base cushion, and secure the parcel shelf to the seat back by pressing and turning the quarter-turn fastener **(2)**. When the seat back is returned to its normal position, ensure that it locks securely, and that the side seat belts are not trapped.

To gain maximum luggage space, open the tailgate and disengage the seat belts from the seat by pulling them through the gap between the seat back and base cushions, then slide the front seats forwards to make room. Fold the seat back forwards on to the base cushion (see above), then fold the whole seat forwards. When the seat is returned to its normal position, ensure that it locks securely, with the seat belts routed correctly between the seat base and back cushions, and check that the webbing is not trapped or damaged.

If the split seat is fitted, extra luggage can be accommodated while retaining at least some

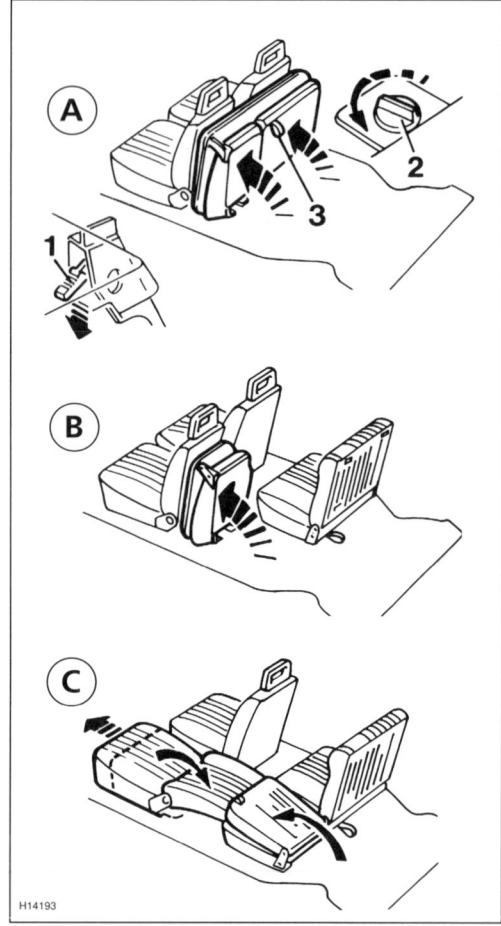

H14193

▲ *Rear seat folding – Maestro*
- **A** *Maximum luggage space*
- **B** *Part load space*
- **C** *Extra-long loads*
- **1** *Seat back locking lever*
- **2** *Quarter-turn fastener*
- **3** *Seat back strap*

rear seat passenger capacity. Remove the parcel shelf, release the seat belt(s) and the appropriate locking lever, then fold forwards the section of seat required, either on to its base cushion or fully forwards, as desired.

To carry long articles, fold the rear seat left-hand section on to its base cushion, then push the front passenger seat fully forwards and recline the seat back; use the seat belts if necessary to strap the load in, preventing any risk of the load falling across the controls while the car is being driven.

MONTEGO SALOON

Where a rear seat with a split back is fitted, slide the front seats forwards to make room. Fold up the centre armrest (where applicable), then open the boot and pull the appropriate release handle to fold forwards the section of the back rest required. Ensure that the seat belts which are not affected are in position for use.

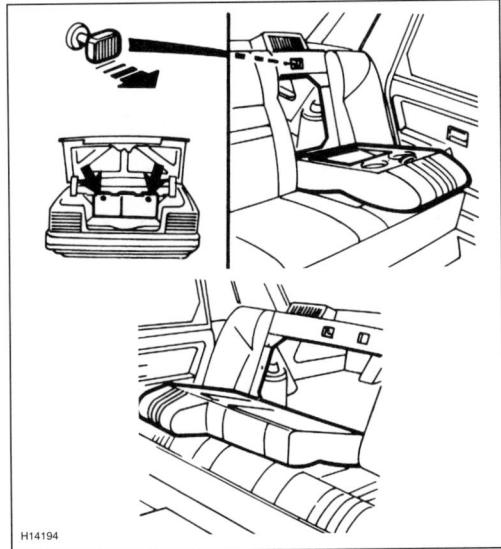

▲ *Rear seat folding – Montego Saloon*

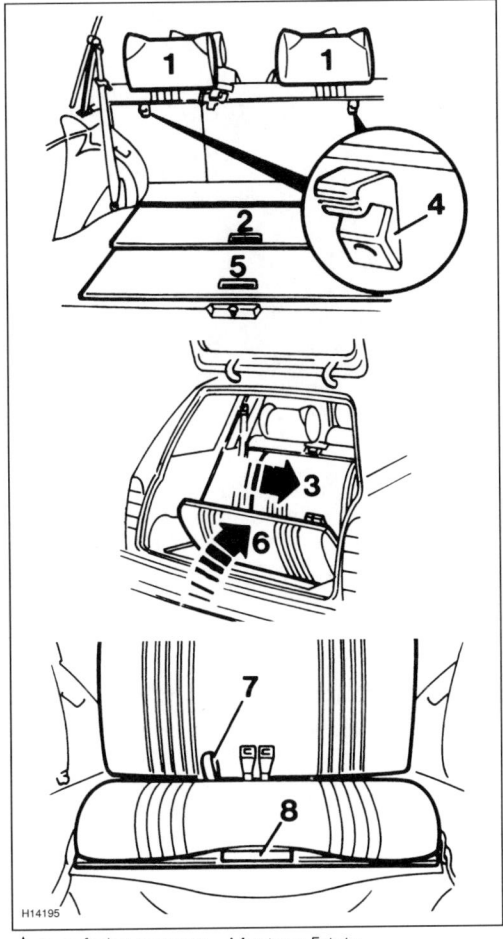

▲ *Rear-facing rear seat – Montego Estate*
1 *Forward-facing rear seat head restraints*
2 *Seat back rest handle*
3 *Seat back rest – raising*
4 *Seat back rest catches – on rear of forward-facing seat*
5 *Seat base cushion handle*
6 *Seat base cushion – raising*
7 *Seat base loop*
8 *Safety warning label*

When refitting, ensure that the back rest lock is securely fastened by trying to pull it forwards.

MONTEGO ESTATE

Where the rear-facing rear seat is fitted, it is for use **only** by (no more than two) children; the weight of each child must **not** exceed 45 kg (7 stone). The seat must be used **only** when the forward-facing seat is erected and its head restraints **(1)** are in place; note that children under the age of 7 may be too small for the seat belts provided – see a Rover dealer for advice, if in doubt.

To erect the rear-facing rear seat, remove the luggage compartment cover (see below), fold back the luggage compartment floor carpet, then raise the seat back rest **(3)** and engage its upper corners in the catches **(4)** on the rear of the forward-facing seat. Grasping the handle **(5)**, raise the seat base cushion **(6)** and lower it into position; always follow the instructions on

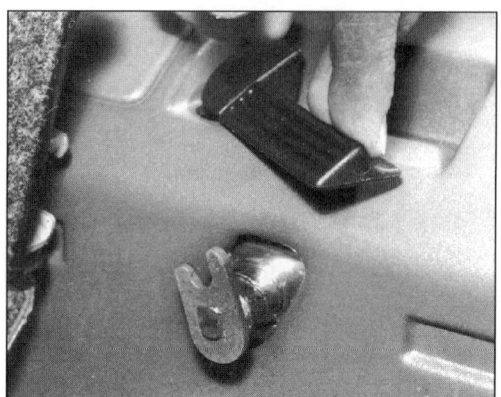

▲ *Releasing forward-facing rear seat back rest lock –
Montego Estate*

the safety warning label **(8)** attached to the
right-hand side channel of the base cushion.

To fold away the seat, use the loop **(7)** to
pull up the base cushion, and tuck the seat
belts into their recesses; otherwise, reverse the
erecting procedure.

To gain maximum luggage space, remove the
luggage compartment cover and fold away the
rear-facing rear seat (as applicable), then open
one rear door and check that the seat belt locks
and fittings are stowed out of the way, in their
pockets (where fitted). The head restraints
(where fitted) should be removed and stowed
in the rearmost compartment under the
luggage compartment floor. Using the loops

provided, pull the seat base cushion(s) forwards
until they are upright behind the front seats;
slide the front seats forwards to make room if
required. Release the back rest locks by turning
them inwards and stow the centre seat belt,
then fold the back rest(s) forwards.

When the seat is returned to its normal
position, ensure that it locks securely, with the
seat belts routed correctly, and check that the
webbing is not trapped or damaged.

If the split seat is fitted, extra luggage can be
accommodated while retaining at least some
rear seat passenger capacity. Release the seat
belt(s) and the appropriate locking lever, then
fold forwards the section of seat required.

Rear parcel shelf

On models with a rear parcel shelf, the shelf
can be removed by unhooking the support
straps **(1)** from the tailgate, and pushing the
shelf to the right until the pivot pin is clear of
the left-hand hinge **(3)**.

Luggage compartment cover

EARLY MODELS

The cover is stowed in a bag in the rearmost
compartment under the luggage compartment
floor; the rear seats must be upright to provide
an enclosed area. Unpack the cover and,
starting at the front, engage the fasteners
along one side. Pull the cover taut and attach
the remaining side.

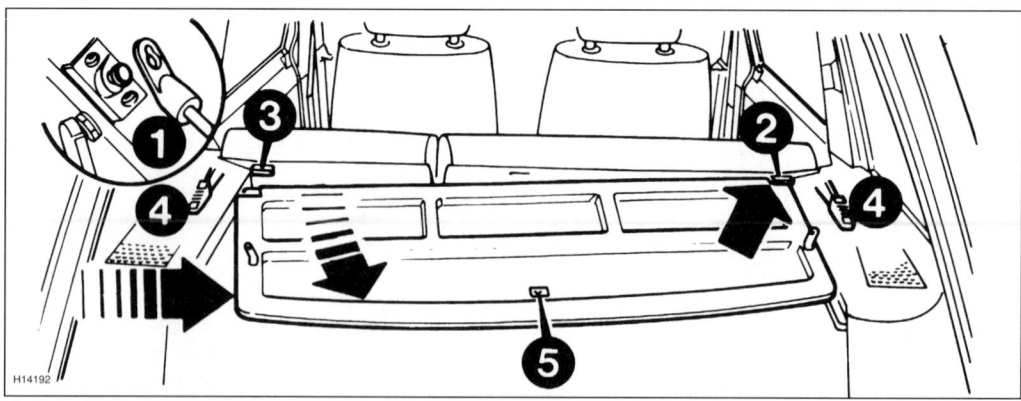

▲ *Removing the parcel shelf – Maestro*

1 Support strap	**3** Left-hand hinge	**5** Quarter-turn fastener
2 Right-hand hinge	**4** Seat back locking levers	

To remove the cover, release the fasteners by pushing from below – do not pull on the cover. Roll up the cover and stow it in its bag.

LATER MODELS

The roller-blind type of cover fitted to later Montego Estates can be unrolled and fastened at its rear edge by means of the fasteners. The small front cover can then be pulled forwards until its loops can be secured; again the rear seats must be upright to provide an enclosed area.

To remove the cover, release the front cover fasteners, followed by those at the rear, allow the main cover to retract, then pull inwards simultaneously on both end cords and lift out the cover.

Stowage boxes

Montego Estate models have two stowage boxes provided for small articles, one in each rear corner of the luggage compartment. To open the cover, turn the fastener through one-quarter of a turn, and grasp the handle to lift away the cover.

EXTERIOR EQUIPMENT

Door locks

To lock a front door, either close it and turn the key in the lock (towards the front of the car) or, with the door open, depress the lock interior plunger (located at the top of the door, below the window) while holding up the exterior handle, then release the exterior handle and close the door. Note, however, that if you use this latter method, there is a danger that you might lock the keys in the car! To unlock a front door, either turn the key towards the rear of the car, or raise the interior plunger.

To lock a rear door (door open or closed), depress the lock interior plunger; lift the plunger to unlock. The rear door locks incorporate childproof catches, operated by pushing forwards the small lever on the back edge of the door; the door cannot then be opened from the inside.

To unlock the tailgate, insert the key and turn it clockwise, then release, **remove the key** (to prevent it breaking if the tailgate is slammed shut or any items on the keyring

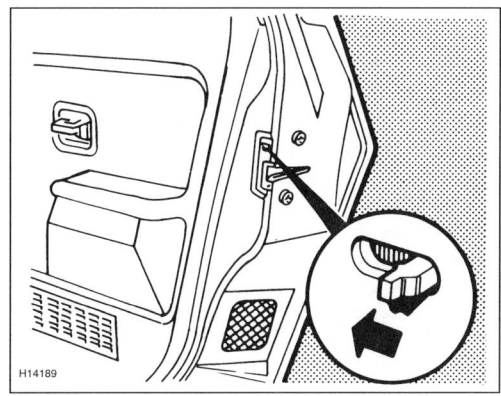

▲ *Operating rear door lock childproof catches*

damaging the paintwork), and depress the locking button to release the latch. To lock the tailgate, close it securely, insert the key and turn it anti-clockwise.

Central door locking system

The system is operated from the driver's door, and enables all other doors and the tailgate to be locked/unlocked electrically whenever the driver's door is locked/unlocked. The tailgate can be operated independently of the system, when required.

Unlocking/locking the driver's door by either of the methods described above will operate the system, but if the key is used, it is essential that you turn the key to the limit of its travel (towards the front of the car to lock, towards the rear to unlock), pause momentarily (to ensure that the system is activated – you will hear the lock motors working), then return the key to the vertical and withdraw it.

From inside the car, operating the driver's door lock interior plunger will lock/unlock all doors, while operating any other door's plunger will act on that door alone.

If a door is unlocked manually once the central locking system has been operated, that door must be relocked manually, as the system will not automatically re-engage.

To open the tailgate when it has been locked manually, proceed as described above.

To open the tailgate when it has been locked by the system, insert the key and turn it half a turn anti-clockwise while applying slight pressure to the locking button, then turn it half a turn clockwise, remove the key and depress the locking button to release the latch and open the tailgate.

To open the tailgate when it has been 'double-locked' (ie, both manually and by the system), insert the key and turn it half a turn clockwise while applying slight pressure to the locking button, then release and remove the key, and depress the locking button to release the latch and open the tailgate.

To lock the tailgate independently of the system, proceed as described above; to lock the tailgate using the system, turn the key half a turn clockwise to the unlocked position before locking the driver's door.

Integral roof rack

The integral roof rack fitted to some Montego Estates must be used **only** with **all** six cross-rails fitted, and with the load securely lashed to the side rails. The load must **not** exceed the maximum permissible (refer to *'Dimensions and weights'* on page 14), must be evenly distributed, and must **not** exceed a height of 200 mm (8 inches) above the lower cross-rails. No more than 25 kg (55 lb) may be supported by any one cross-rail, while long loads supported by the upper cross-rails only must **not** exceed 30 kg (66 lb).

To fit the cross-rails, unload them from their bag in the rearmost compartment under the

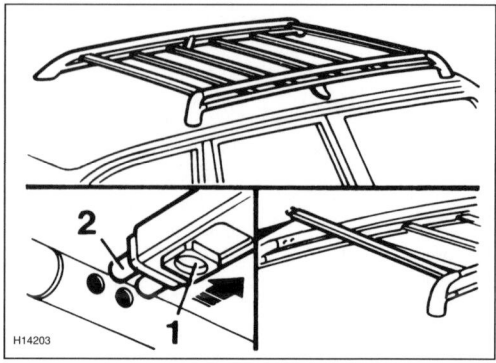

▲ *Integral roof rack details – Montego Estate*
  1 *Button*
  2 *Spring-loaded locating pins*

luggage compartment floor. Noting that the shorter cross-rails fit between the upper side rails, engage each cross-rail's fixed locating pins in turn in one side rail, then slide the button **(1)** in the cross-rail's opposite end inwards to allow the corresponding spring-loaded locating pins **(2)** to enter into the side rail; release the button and check that the cross-rail is securely fastened.

To remove the cross-rails, reverse the removal procedure; use the roadwheel trim removal tool (refer to *'Breakdowns'* on page 51), if required, to pull back the button.

In the event of an accident, the first priority is safety. This may seem obvious, but in the heat of the moment, it's very easy to overlook certain points which may worsen the situation, or even cause another accident. The course of action to be taken will vary depending on how serious the accident is, and whether anyone is injured, but always try to think clearly, and don't panic.

We don't suggest that you consult this Section at the scene of an accident, but hopefully the following advice will help you to be better prepared to deal with the situation should you be unfortunate enough to appear on the scene of an accident or become involved in one yourself.

HOW TO COPE WITH AN ACCIDENT

Deal with any possible further danger

Further collisions and fire are the dangers in a road accident.

● **If possible warn other traffic**

Where possible, switch on the car's hazard warning flashers.

If a warning triangle is carried, position it a reasonable distance away from the scene of the accident, to give approaching drivers sufficient warning to enable them to slow down. Decide from which direction the approaching traffic will have least warning, and position the triangle accordingly.

If possible, send someone to warn approaching traffic of the danger which exists, and to encourage the traffic to slow down.

● **Switch off the ignition and impose a 'No Smoking' ban**

This will reduce the possibility of fire, should there be a petrol leak.

Call for assistance

Send someone to call the emergency services (dial 999), and make sure that all the necessary information is provided to the operator. Give the exact location of the accident, and the number of vehicles and if applicable the number of casualties involved. Refer to the *'Motorway breakdowns'* Section on page 48 for details of how to call for assistance on a motorway.

● **Call an ambulance**

If anyone is seriously injured or trapped in a vehicle.

● **Call the fire brigade**

If anyone is trapped in a vehicle, or if you think that there is a risk of fire.

● **Call the police**

If any of the above conditions apply, or if the accident is a hazard to other traffic. In most cases, the accident must be reported to the police within 24 hours even if none of the above conditions apply (refer to *'Requirements of the law in the event of an accident'* on page 42).

Administer first aid

Refer to *'First aid'* on page 40.

Provide your details

If you are involved in the accident as a driver or car owner, provide your personal and vehicle details to anyone having reasonable grounds to ask for them. Also inform the police of the details of the accident as soon as possible, if not already done.

ACCIDENTS

FIRST AID

Always carry a first aid kit

If possible, learn first aid by attending a suitable course – contact the St John Ambulance Association or Brigade, St Andrew's Ambulance Association or the British Red Cross Society in your local area for details. These organisations will also be able to provide further training in heart massage and mouth-to-mouth resuscitation which could enable you to save someone's life in an emergency.

Before proceeding with any kind of first aid treatment, deal with any possible further danger, and call for assistance, as described previously in this Chapter.

To help remember the sequence of action which should be followed when dealing with a seriously injured casualty, use the **ABC** of emergency first aid.

Casualties remaining in vehicles

Any injured people remaining in vehicles should not be moved unless there is a risk of further danger (such as from fire, further collisions etc).

Casualties outside the vehicles

Recovery position

To prevent the possibility of an unconscious but breathing casualty choking or suffocating, he/she must be placed in the recovery position.

Lie the casualty on his/her stomach with the face to one side and the knee and arm bent as shown.

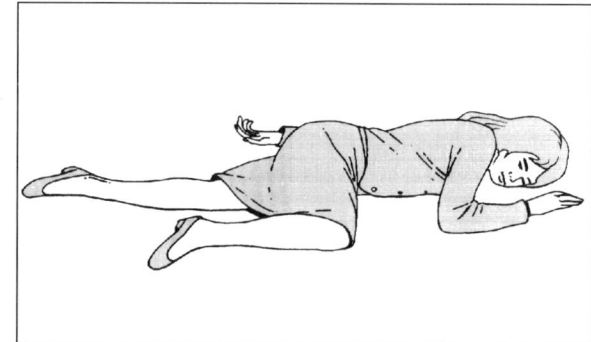

▲ *The recovery position*

First aid for other injuries

Always treat injuries in the following order of priority:
- Ensure that the casualty has a clear airway. If not treat as described opposite.
- Put any breathing, unconscious casualty into the recovery position as described previously.
- Treat any severe bleeding by applying direct pressure to the wound. Maintain the pressure for at least 10 minutes and apply a suitable clean dressing. If the wound continues to bleed, apply further pressure and dressings **over** that already in place. If bleeding from a limb, and as long as the limb is not broken, lift the affected limb to reduce the bleeding.
- Broken limbs should not be moved, unless the casualty has to be moved in order to avoid further injury. Support the affected limb(s) by placing blankets, bags, etc, alongside.
- Burns should be treated with plenty of cold water as soon as possible. Don't attempt to remove any clothing, but cover with a clean dressing.

Reassurance

The casualty may be in shock, but prompt treatment will minimise this. Reassure the casualty confidently, avoid unnecessary movement, and keep him/her comfortable and warm. Make sure that the casualty is not left alone.

Give the casualty NOTHING to eat, drink or smoke.

Airway
● Check for breathing
Check that the airway is clear. If the casualty is breathing noisily or appears not to be breathing at all, remove any obvious obstruction in the mouth and tilt the head back as far as possible. Breathing may then start.

Breathing
● If breathing has stopped
If after clearing the airway the casualty still appears not to be breathing, look to see if the chest or abdomen is moving. Place your ear close to the casualty's mouth to listen and feel for breathing, and look to see if there's any movement of the chest. If you can't detect anything, you must start breathing for the casualty (mouth-to-mouth resuscitation).

To do this, pinch the casualty's nostrils firmly, keep his/her chin raised, and seal your lips around the casualty's mouth. Breathe out through your mouth into the casualty until the chest rises, then remove your mouth and allow the casualty's chest to fall. Give one further breath.

If giving mouth-to-mouth resuscitation to a child, remember that an adult's lungs are significantly larger than those of a child. Care must be taken not to overinflate a child's lungs, as this may cause injury.

● Check for a pulse at the base of the neck
If the heart is beating, continue to breathe into the casualty at a rate of one breath every 5 seconds, until he/she is able to breathe unaided, then place the casualty in the recovery position (refer to 'Recovery position' opposite).

Circulation
● If the heart has stopped
If there is no pulse at the neck, then external chest compression (heart massage) must be started.

To do this, if the casualty is still in the vehicle, he/she must be removed and laid on his/her back on the ground. Kneel down on one side of the casualty. Feel for the lower half of the casualty's breastbone and place the heel of one of your hands on this part of the bone, keeping your fingers off the casualty's chest. Cover this hand with your other hand, interlocking your fingers. Keeping your arms straight and vertical, press the breastbone down about 4 to 5 centimetres (1½ to 2 inches). The pressure should be smooth but not jerky. Do this 15 times, at the rate of just over once a second. It may help you to count aloud as you press.

Follow this by 2 breaths as described under 'Breathing'.

Repeat the cycle of 2 breaths followed by 15 compressions, twice more, ending with 2 breaths, then re-check the pulse.

● If there is still no pulse
Repeat the cycle 9 times and then check the pulse. Continue this procedure until there is a pulse or until professional help arrives.

REQUIREMENTS OF THE LAW IN THE EVENT OF AN ACCIDENT

The following is taken from the Road Traffic Act of 1988.

If you are involved in an accident – which causes damage or injury to any other person, or another vehicle, or any animal (horse, cattle, ass, mule, sheep, pig, goat or dog) not in your vehicle, or roadside property:

You must
- stop;
- give your own and the vehicle owner's name and address and the registration mark of the vehicle to anyone having reasonable grounds for requiring them;
- if you do not give your name and address to any such person at the time, report the accident to the police as soon as reasonably practicable, and in any case within 24 hours;
- if anyone is injured and you do not produce your certificate of insurance at the time to the police or to anyone who has with reasonable grounds required its production, report the accident to the police as soon as possible, and in any case within 24 hours, and either produce your certificate of insurance to the police when reporting the accident or ensure that it is produced within seven days thereafter at any police station you select.

ESSENTIAL DETAILS TO RECORD AFTER AN ACCIDENT

If you're involved in an accident, note down the following details which will help you to complete the accident report form for your insurance company, and will help you if the police become involved.
- The name and address of the other driver and those of the vehicle owner, if different.
- The name(s) and address(es) of any witness(es) (independent witnesses are particularly important).
- A description of any injury to yourself or others.
- Details of any damage caused to the vehicles involved or other property.
- The name and address of the other driver's insurance company and, if possible, the number of his/her certificate of insurance.
- The registration number of the other vehicle (check this against the tax disc if possible).
- The number of any police officer attending the scene.
- The location, time and date of the accident.
- The speed of the vehicles involved.
- The width of the road, details of road markings and signs, the state of the road surface, and the weather conditions.
- Any marks or debris on the road relevant to the accident.
- A rough sketch showing the vehicle positions before and after the accident. It's helpful to make a note of the vehicle positions in terms of distance from fixed landmarks, such as lamp posts, buildings, etc.
- Whether any of the other vehicle occupants were wearing seat belts.
- If the accident occurred at night or in poor visibility, whether vehicle lights or street lights were switched on.
- If you have a camera, take a picture of the vehicles and the scene.

If the other driver refuses to give you his/her name and address, or if you consider that he/she has committed a criminal offence, inform the police immediately.

Use this page to record all the relevant details in the event of an accident.
You should transfer all the information recorded on this page onto the Motor Vehicle Accident Report form which you can obtain from your insurance company.

OTHER DRIVER DETAILS

FULL NAME OF DRIVER:

ADDRESS:

POST CODE:

HOME TELEPHONE: WORK TELEPHONE:

DRIVING LICENCE NUMBER: ISSUED:

DATE OF BIRTH: DATE DRIVING TEST PASSED:

TYPE OF LICENCE: (tick box) Full ☐ Provisional ☐ Heavy Goods ☐

PERMITTED GROUPS:

FULL NAME OF OWNER:

ADDRESS:

POST CODE:

TELEPHONE:

INSURANCE COMPANY:

POLICY NUMBER:

AGENT OR BROKER:

TELEPHONE:

OTHER VEHICLE DETAILS

MAKE: MODEL:

YEAR: CC:

REGISTRATION NUMBER:

CIRCUMSTANCES OF ACCIDENT

DATE: TIME: am/pm

PLACE: (Street or Road)

TOWN: COUNTY:

SPEED: WERE THE POLICE CALLED? Yes ☐ No ☐

If 'Yes' give details of Police Constabulary concerned:

DETAILS OF WHAT HAPPENED (Please use the page opposite to make a rough sketch map)

WHAT WAS THE WEATHER LIKE?

INDEPENDENT WITNESS 1

TELEPHONE:

INDEPENDENT WITNESS 2

TELEPHONE:

DETAILS OF ANY INJURIES SUSTAINED BY EITHER PARTY:

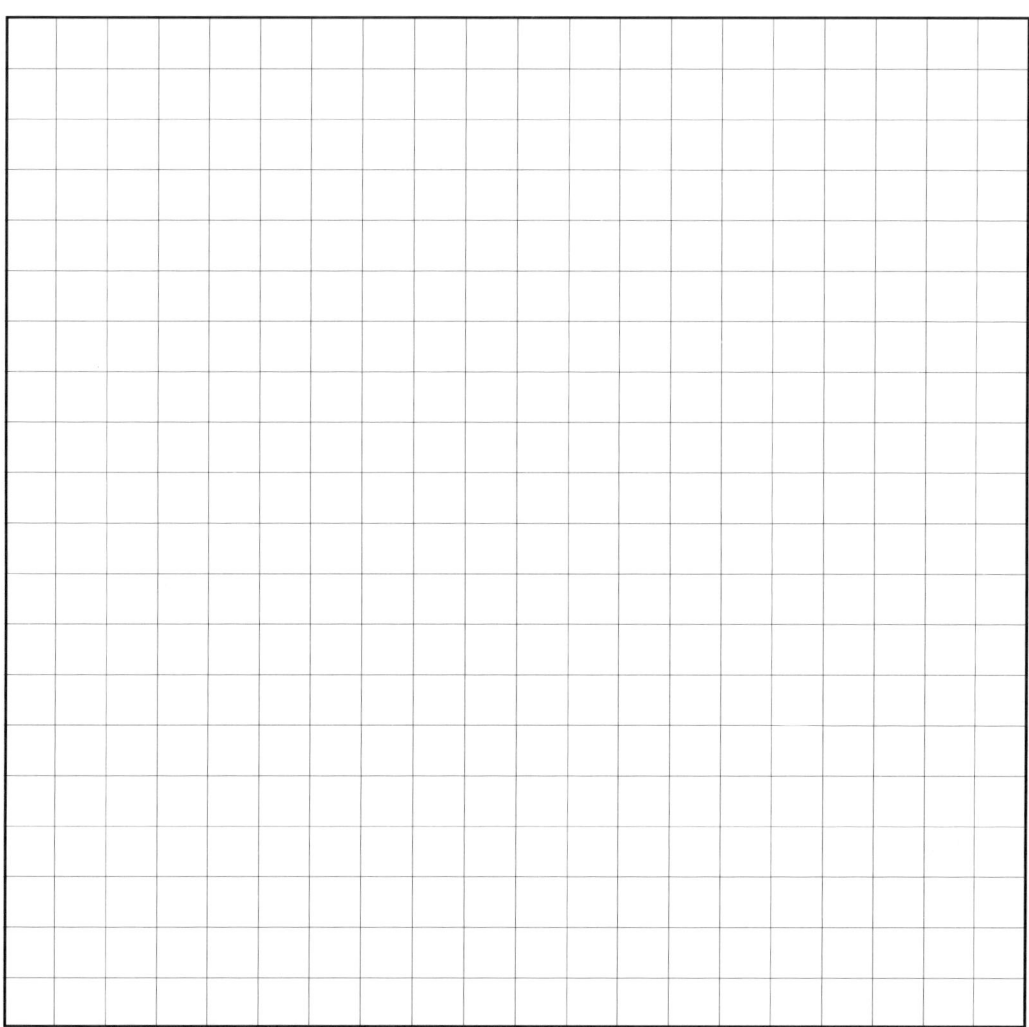

Essential items to include in your sketch map:
- The layout of the road and its approaches.
- The directions and identities of both vehicles.
- Their relative positions at the time of impact.
- The road signs and road markings.
- Names of the Streets and Roads.

HOW TO COPE WITH A FIRE

Fire is unpredictable, and particularly dangerous when cars are involved because of the presence of petrol, which is highly inflammable.

It's sensible to carry a fire extinguisher on board your car, but bear in mind that the average car fire extinguisher is suitable for use only on the smallest of fires.

In the event of your car catching fire:

- **Switch off the ignition.**
- **Get any passengers and yourself out of the car and well away from danger.**
- **Impose an immediate 'No Smoking' ban.**
- **DO NOT expose yourself, or anyone else, to unnecessary risk in an attempt to control the fire.** Minor fires can be controlled using a suitable extinguisher, but always use an extinguisher at arms-length, and bear in mind that a small vehicle fire can develop into a serious situation without warning.
- **Call for assistance, if necessary** – refer to *'How to cope with an accident'* on page 39.

HOW TO COPE WITH A BROKEN WINDSCREEN

There are a number of national specialist companies who offer roadside assistance to drivers who suffer broken windscreens, and it's a good idea to carry the 'phone number of a suitable specialist in your car for use in such situations. Some car insurance policies enable the use of such companies at preferential rates, and it's worth enquiring about this when obtaining insurance quotes.

- **Most of the windscreens fitted to modern cars are of the laminated type.** Laminated windscreens are made from layers of glass and plastic (usually two layers of glass sandwiching a layer of plastic) which prevents the windscreen from shattering, and preserves the driver's vision in the event of the screen suffering an impact.
  A sharp impact may crack the outer glass layer, but clear vision is usually maintained.

Obviously, if the windscreen is damaged, it should be replaced at the earliest opportunity, but there is no need to curtail your journey unless the damage is particularly serious.

- **Some older cars may be fitted with toughened windscreens.** If a toughened windscreen suffers an impact, the glass will normally stay intact, but severe 'crazing' will usually occur which is likely to seriously affect the driver's vision.
  If a toughened windscreen breaks, stop immediately, and seek assistance. **DO NOT** attempt to knock the broken glass out of the windscreen frame, as this is likely to result in injury to yourself, and damage to the car. Driving without a windscreen is extremely dangerous and should not be attempted.

WHAT TO DO IF YOUR CAR IS BROKEN INTO OR VANDALISED

If you are unfortunate enough to have your car broken into or vandalised, do not attempt to drive your car away from the scene unless you are satisfied that no damage has been caused which could affect the car's safety.

Where possible notify the police before moving your car, and inform your insurance company at the earliest opportunity.

In the unfortunate event of a breakdown, the first priority must always be safety – never risk causing an accident by attempting to move the car or trying to work on it if it's in a dangerous position.

Joining one of the national motoring organisations such as the AA or the RAC can provide you with a recovery and assistance service should you break down. It may also save you money, as local garages often charge high rates to provide a recovery service.

This Section provides advice on the course of action to follow should you break down. There are also Sections covering the procedure to follow for towing, and on how to cope with two of the more common causes of breakdowns – a puncture, and a flat battery.

Breakdowns on an ordinary road (not a motorway)

Most importantly of all, common sense must be used, but the following advice should prove helpful.

- **Warn approaching traffic,** to minimise the risk of a collision.

    Where possible, switch on the car's hazard warning flashers, and at night leave the sidelamps switched on.

    If a warning triangle is carried, position it a reasonable distance away from the scene of the breakdown, to give approaching drivers sufficient warning to enable them to slow down. Decide from which direction the approaching traffic will have least warning, and position the triangle accordingly. Place the warning triangle on the road surface, out from the edge of the road, where it can be easily seen.

    If possible, send someone to warn approaching traffic of the danger which exists, and to encourage the traffic to slow down.
- **Get the passengers out of the car** as a precaution should the car be hit by another vehicle. The passengers should move well away from the car and approaching traffic, up (or down) an embankment, or into a nearby field for example.
- **If possible, move the car to a safe place.** If the car cannot be safely driven or pushed to a place of safety, call a recovery service to provide assistance. *Don't expose yourself or anyone else to unnecessary danger in order to move the car.*
- **If the car can't be safely moved, turn the steering wheel towards the side of the road.** If the car is hit, it will then be pushed into the side of the road, not into the path of traffic.
- **Try to find the cause of the problem.** Only attempt this if the car is in a safe position away from traffic. Refer to *'Fault finding'* on page 139 for details.
- **If necessary, call for assistance** from one of the national motoring organisations if you're a member, or from a local garage who can provide a recovery service.

Motorway breakdowns

In the event of a motorway breakdown, due to the speed and amount of traffic there are a few special points to consider, and the following advice should be followed.

- **If possible, switch on the car's hazard warning flashers, and at night leave the sidelamps switched on.**
- **DO NOT open the doors nearest to the carriageway, and DO NOT stand at the rear of the car, or between it and the passing traffic.**
- **Get the car off the carriageway and onto the hard shoulder as quickly as possible, and as far to the left as possible** – never forget the danger from passing traffic. Turn the steering wheel to the left so that the car will be pushed away from the carriageway if hit from behind.
- **If you can't move the car off the carriageway** – get everyone out of the car and well away from the carriageways (up/down an embankment, or into a nearby field for example) as quickly as possible. Never forget the danger from passing traffic.

    Call for assistance as described below.

    Attempt to warn approaching traffic of your car's presence by signalling from well to the back of the hard shoulder, NOT the carriageway. DO NOT put yourself at risk. Face the approaching traffic, and be prepared to move quickly clear of the hard shoulder onto the verge if any vehicles pull onto the hard shoulder.

● **Get the passengers out of the car** – as a precaution should the car be hit by another vehicle. The passengers should move well away from the car and approaching traffic, up (or down) an embankment, or into a nearby field for example. If animals are being carried, it may be advisable to leave them in the vehicle. In any case, animals and children **must** be kept under tight control.

● **Call for assistance** – using the nearest emergency telephone on your side of the carriageway. NEVER cross the carriageways to use the emergency telephones.

The direction of the nearest telephone is indicated by an arrow on the marker post behind the hard shoulder (the marker posts are positioned 100 metres/300 feet apart).

Each telephone is coded with details of its location, and the operator will automatically be informed of your position on the motorway when you make the call.

Give the operator details of your car type and registration number and, if you're a member of one of the motoring organisations, ask the operator to arrange assistance.

● **Don't leave your car unattended for a long period.**

Changing a wheel

● **First of all ensure that the car is in a safe position.** It may be better to drive the car slowly forwards to a safer location on firm, level ground and risk damaging the wheel, than to risk causing an accident.

● **If you're in any doubt** as to whether you can safely change the wheel without putting yourself or other people at risk, call for assistance.

● **Apply firmly the handbrake** and select first or reverse gear (manual gearbox) or position **P** (automatic transmission).

● **Ask any passengers to get out of the car** and to stay clear, then unload any luggage you may be carrying, from the roof rack (if used) as well as from the boot/luggage

compartment; if you are towing a caravan or trailer, unhitch it before jacking up the car.

● **All the tools required to change a wheel are kept, with the spare, under the boot/luggage compartment floor.**

● **On all Maestro models,** lift the luggage compartment floor mat (and, where fitted, the fibreboard cover beneath, noting how it is clipped at its front edge), then lift out of the spare wheel the insert containing the tools and jack.

Using the wheel brace (the L-shaped spanner provided to unscrew the wheel nuts), unscrew the bolt securing the spare wheel's locking plate and remove the spare wheel.

▲ Unscrew locking plate bolt (arrowed) to release spare wheel – Maestro

▲ Tools are kept in spare wheel – Maestro

● **On Montego Saloon models,** lift the boot floor mat (and, where fitted, the fibreboard cover beneath – note how it is clipped at its front edge), then lift out of the spare wheel the insert **(4)** containing the tools **(2, 3)**. Using the wheel brace **(3)**, unscrew the bolt securing the spare wheel's locking plate **(1)**. Slip both loops **(12)** of the sling over the body hooks **(13)** and pull on the sling's single strap **(14)** until its crutch can be

hooked over the boot lid striker; lift off the spare wheel. Using a coin, release the fastener **(11)** by rotating it through one quarter of a turn, then lift the cover **(9)** upwards and sideways to release the clip **(10)**; remove the jack handle **(8)**. Noting how the jack **(5)** is secured by its retainer **(6)**, rotate anti-clockwise the jack screw until the retainer falls clear and the jack can be removed.

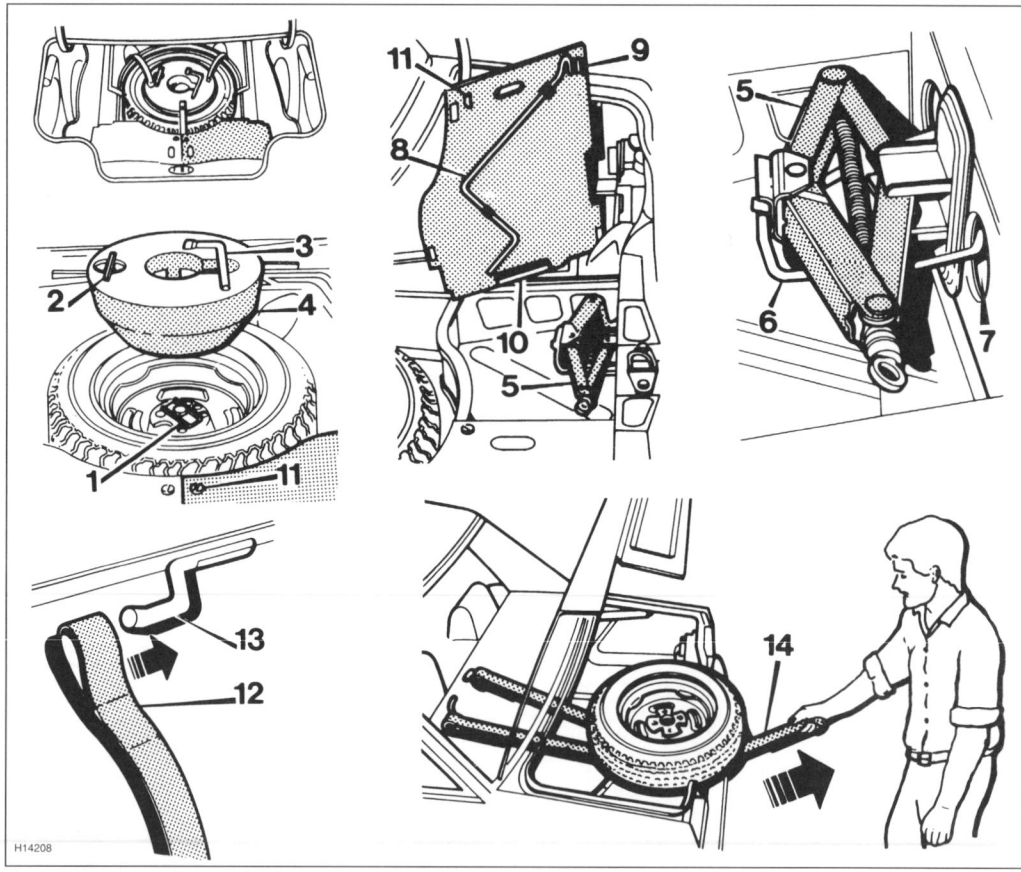

▲ *Tool stowage and spare wheel removal – Montego Saloon*

1	Spare wheel locking plate	**8**	Jack handle
2	Wheel trim removal tool	**9**	Fibreboard cover
3	Wheel brace	**10**	Clip
4	Moulded insert	**11**	Quarter-turn fastener
5	Jack	**12**	Sling loop (one of two)
6	Jack retainer	**13**	Body hook (one of two)
7	Rear body panel cut-out	**14**	Sling single strap

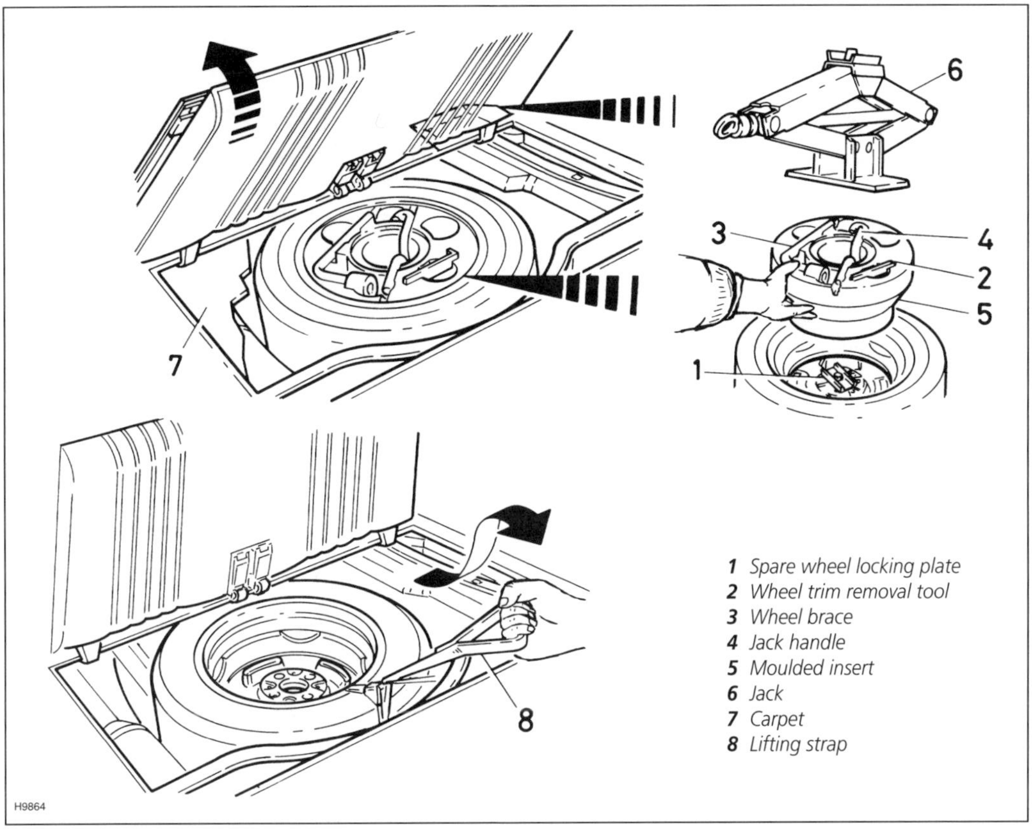

1 Spare wheel locking plate
2 Wheel trim removal tool
3 Wheel brace
4 Jack handle
5 Moulded insert
6 Jack
7 Carpet
8 Lifting strap

H9864

▲ Tool stowage and spare wheel removal – Montego Estate

● **On Montego Estate models,** fold back the luggage compartment floor mat/carpet. Lift the luggage compartment floor or, where fitted, the rear-facing rear seat backrest (7-seater models), and clip it against the front-facing rear seat backrest. Remove the jack **(6)** and, where fitted, lift the carpet **(7)** covering the spare wheel, then lift out of the spare wheel the insert **(5)** containing the tools **(2, 3 and 4)**. Using the wheel brace **(3)**, unscrew the bolt securing the spare wheel's locking plate **(1)**. Using the lifting strap **(8)**, lift out the spare wheel.

● **On all models, when the spare wheel has been removed,** check now that its tyre pressure is correct, to save wasting time by fitting a flat spare!

● **To get at the wheel nuts**, it will be necessary on most models to remove some item of trim. Use the special tool provided to prise off the trim covering either the whole wheel or just the nuts (where applicable)

▲ Prising trim from wheel centre to reach wheel nuts

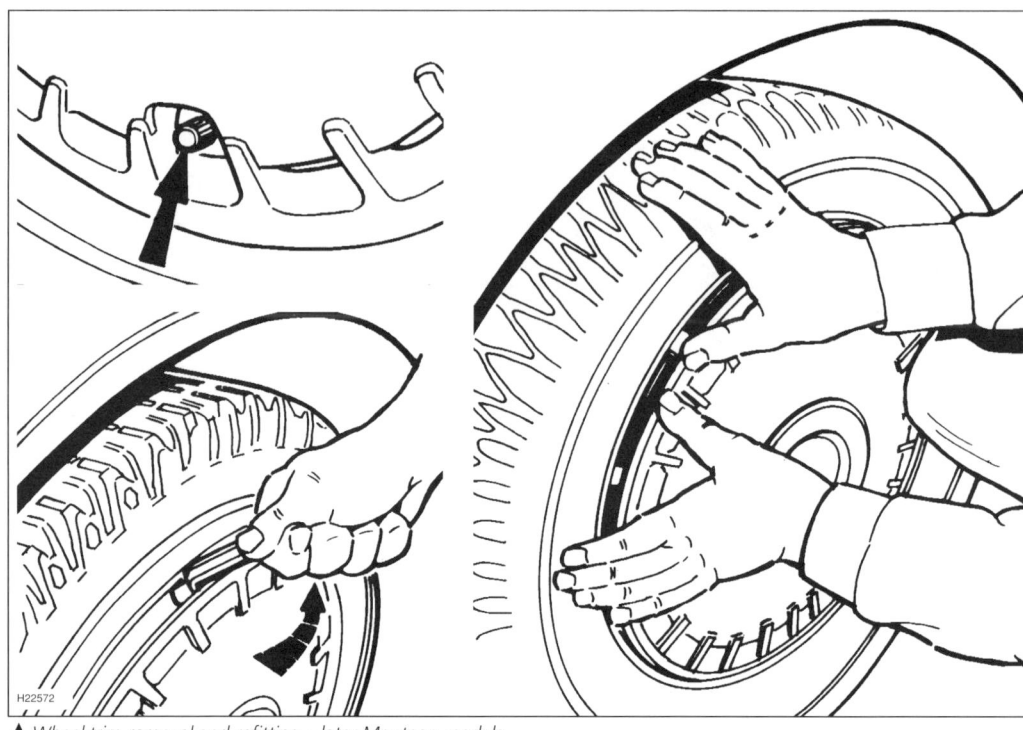

▲ *Wheel trim removal and refitting – later Montego models*

around the wheel's centre. Try not to mark the plastic trim, and do not jerk it sharply off the wheel or you may break the retaining clips.

● **Now you have to slacken the wheel nuts** – this may be the most difficult bit, especially if you have had new tyres fitted and the fitter was over-enthusiastic when tightening the nuts! If all you have with you is the factory-supplied wheel brace, be very careful if the nuts are tight – the brace is awkward to use in these circumstances, and can quite easily slip off the nut. The best answer to such situations is to buy one of the properly-shaped braces with an extending handle that are widely available at most good car accessory shops. If you don't have one of these, the next best solution is to slip a length (about 1 to 2 ft would be enough) of strong tubing over the end of the wheel brace, to give you the extra leverage needed. Slacken the nuts anti-clockwise through half to one turn each,

working in a diagonal sequence (ie, start at the top nut, then the bottom one, the left-hand nut and the right-hand one).

● **To make sure that the car cannot move and slip off the jack,** place chocks (any blocks of wood, bricks or large stones lying around will do) at the front and rear of the wheel diagonally opposite the one to be changed.

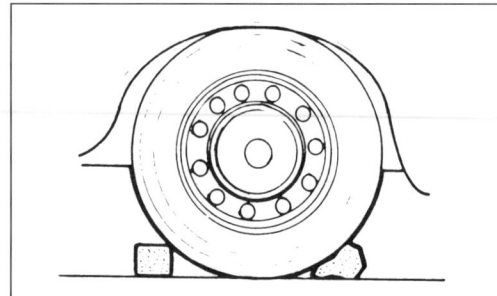

▲ *Chock the diagonally-opposite wheel to prevent the car from moving while jacked up*

● **Locate the jack head's peg in the jacking point** nearest to the wheel to be changed; ensure that the peg on the jack's head fits securely into the hole in the jacking point.

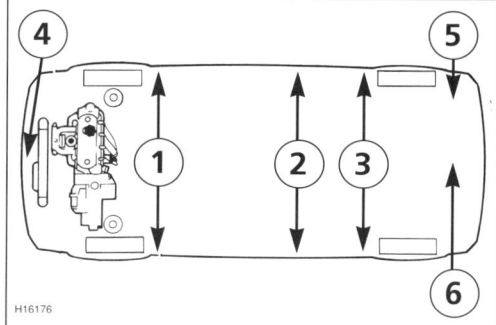

▲ *Location of jacking points for roadside wheel changing (towing eyes also shown)*
  1 Front jacking points – all models
  2 Rear jacking points – Maestro, Montego Saloon
  3 Rear jacking points – Montego Estate models
  4 Front towing eye
  5 Rear towing eye – Maestro
  6 Rear towing eye – Montego

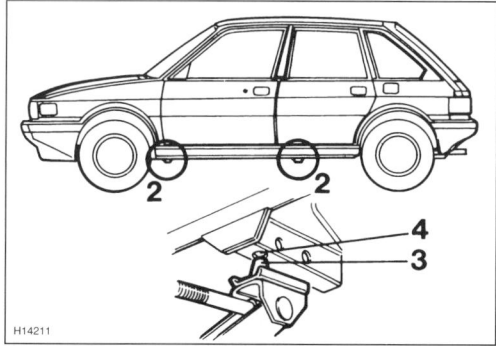

▲ *Locate peg on jack head in jacking point hole*
  2 Jacking points
  3 Jack head peg
  4 Jacking point recess

● **Turn the jack screw clockwise** until the jack extends so that its base plate rests evenly on firm ground; if the ground is likely to be soft, try to find a piece of wood which will spread the load.

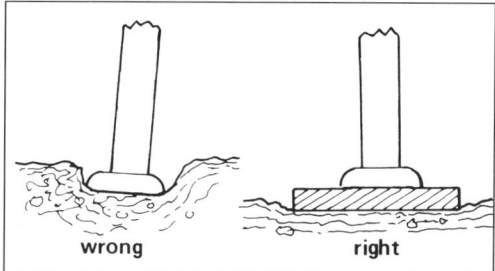

▲ *Using a piece of wood to spread the load under the jack*

● **Jack up the car** by unfolding the jack's handle (Maestro) or by fitting the jack handle's hooked end to the jack eye (Montego); turn the handle clockwise slowly to raise the car until the wheel is clear of the ground, making sure that the car doesn't move on the jack. Slide the spare wheel under the door sill panel for the time being, as near to the punctured wheel as possible. This will break the car's fall (and minimise the risk of injury to you) if the car slips off the jack, or if the jack collapses.

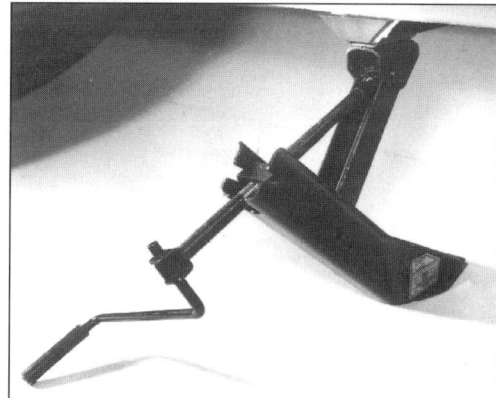

▲ *Maestro jack in use*

● **When the wheel is clear of the ground,** remove the nuts and lift off the wheel; if it is stuck to the hub, give it a good thump from behind to free it, but be careful not to move the car on the jack. Remove the spare wheel from under the side of the car, and slide the punctured wheel under there in its place for the time being.

- **Check that the wheel retaining stud threads** and the wheel-to-hub mating surfaces are clean and undamaged. The threads should be cleaned with a brass wire brush if rusty and a thin smear of copper-based anti-seize compound (available from any motor accessory shop; high melting-point grease will do if nothing else is available) should be applied to the threads and (especially if light alloy wheels are fitted) to the wheel-to-hub mating surfaces to prevent the formation of corrosion. *Obviously, you are not likely to be able to do anything about this by the roadside, but remember to have the work done as soon as possible once the car is running again, or it may be even more difficult to change the wheel next time!*

- **Clean the inside of the spare wheel** and transfer any items of trim, then fit the wheel over the wheel studs and refit the nuts. On some cars with light alloy wheels, the nuts will fit only one way; ensure that the wheel is seated correctly on the shoulder of such nuts before tightening them. In most cases, however, the nuts must be fitted with their tapered ends towards the wheel, to engage with the wheel's dished section. Tighten the nuts evenly until the wheel is held securely against the hub.

- **Remove the punctured wheel from under the car and lower the jack, then tighten the nuts securely,** working in progressive stages (ie, by a turn at a time) and in a diagonal sequence – there is no need to strain yourself doing this, just make sure that the nuts are tight, using the wheel brace only (without an extending tube).

- **Refit the wheel trim(s)** (where applicable), using firm hand pressure to seat the clips; on models with plastic trims covering the whole wheel (but particularly later Montegos), position the wheel with the tyre valve stem at the bottom, locate the trim's large cut-out over the valve and press the upper part of the trim into place by applying equal pressure to both sides. Remove the wheel chocks and stow the punctured wheel, the jack and the tools.

- **Make a final check** to ensure that all tools and debris have been cleared from the roadside.

- **Make sure that neutral is selected** before starting up and driving away.

- **Remember to do the following as soon as possible** to ensure that the car is in a safe condition to drive, and to minimise the inconvenience and struggle if you should ever have to change a wheel again by the roadside:

  1 Check the tyre pressure; this **must** be done at the first available opportunity.

  2 Also at the **first** available opportunity, have the tightness of the wheel nuts checked by someone suitably qualified.

  3 Have the spare repaired or renewed, as necessary.

  4 Remember that it may be necessary to have the newly-fitted wheel balanced, especially if it has been fitted to the front of the car.

  5 If the wheel was difficult to remove, take it off again when you get home, clean off any rust or dirt with a wire brush, wipe clean the inside of the wheel and apply a thin smear of lubricant as described above, before refitting the wheel. Repeat this exercise on the remaining wheels in turn; if one was difficult to remove, so will the others be.

  6 If light alloy wheels are fitted, use a torque wrench to check that the wheel nuts are still correctly tightened 600 miles (1000 km) later, whenever a wheel is removed and refitted.

  7 List everything you didn't have at the roadside, but would like to have had (a torch, wheel brace extension, chocks, an old blanket or groundsheet to kneel on, gloves/hand cleaner, etc), and keep it in the car so that you are better-prepared next time!

Towing

If your car needs to be towed, a special towing eye is provided at the front centre of the car, just below and to the rear of the front bumper.

If the car is fitted with a deep front spoiler (eg, MG models), remove the spoiler centre section (secured by a single screw); if the Rover Body Styling kit or similar is fitted, it may be necessary to remove the front bumper/spoiler, or at least to protect it from the tow-bar/rope, before the car can be towed.

The rear towing eye/bracket may be used, in emergency, for towing a light vehicle **only**.

▲ *Front towing eye – all models*

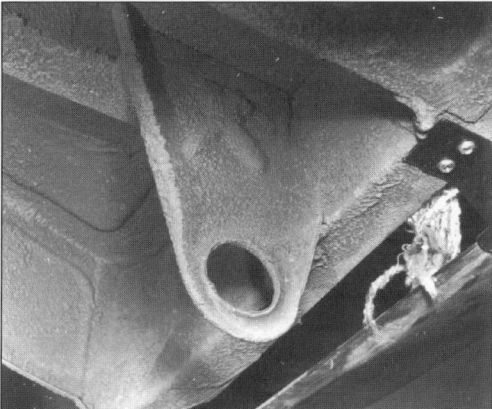
▲ *Rear transporter lashing point/towing eye – Maestro*

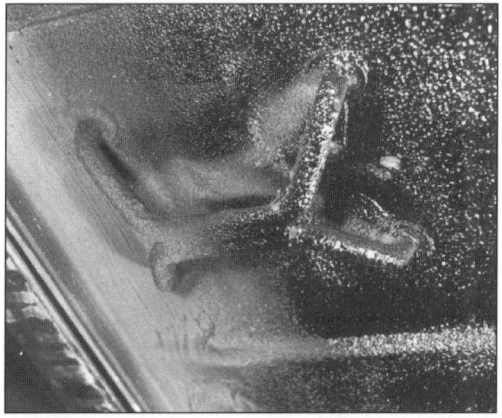
▲ *Rear towing bracket – Montego*

On all models, attach the tow-rope so that when it is pulled taut, it cannot damage, bend, or rub against any suspension components, body panels or the exhaust system.

Never attach a tow-rope to any other part of the body, or to any part of the steering or suspension systems.

If the car (manual gearbox or automatic transmission) is thought to have a transmission fault, **do not** tow it on all four wheels; call for assistance.

If the car is fitted with automatic transmission it should be towed, if possible, only with the front wheels lifted off the road (to avoid serious transmission damage), and should never be towed with the rear wheels alone off the ground. If the car has to be towed with all four wheels on the ground, to avoid transmission damage, do not tow it any faster than 30 mph (50 km/h) or any further than 30 miles (50 km). Before starting, check that the transmission fluid level is correct (refer to *'Servicing'* on page 113).

When being towed, ensure that neutral (manual gearbox) or position **N** (automatic transmission) is selected. Also, the ignition key should be turned to position **II** (steering lock released and ignition warning lamp on). This is necessary for the direction indicators, horn and brake lamps to work.

Note that when being towed, the brake servo does not work (because the engine is not running). This means that the brake pedal will have to be pressed harder than usual to operate the brakes, so allowance should be made for greater stopping distances. If the breakdown doesn't prevent the engine from running, the engine could be started and allowed to idle so that the servo operates while towing. Similarly, on cars fitted with power-assisted steering, the steering will feel much heavier than usual, unless the engine can be started and allowed to idle so that the steering pump can operate.

An 'On Tow' notice should be displayed prominently at the rear of the towed car, to warn other drivers.

Make sure that both drivers know details of the route to be taken before moving off, as it is difficult, and dangerous, to try to communicate once under way.

Before moving away, the towing car should

be driven slowly forwards to take up any slack in the tow rope.

The driver of the towed car should make every effort to keep the tow rope tight at all times, by gently using the brakes if necessary, particularly when driving downhill.

Drive smoothly at all times, particularly when moving away from a standstill. Allow plenty of time to slow down and stop when approaching junctions and traffic queues.

Starting a car with a flat battery

Apart from old age (most batteries should last for at least three years), a battery will normally only go flat if there is a fault in the charging circuit (indicated by a continuously-glowing ignition warning lamp), or when a particular circuit (eg headlamps) is left on for a long time with the engine switched off.

To start a car with a flat battery, you can use a set of jump leads; alternatively, on models with a manual gearbox and no catalytic converter, you could use a push or tow start.

TOW/PUSH STARTING

Warning: *Do not tow/push start any car fitted with a catalytic converter – there is a risk of unburned petrol damaging the converter, or even exploding inside it.*

The method used to start the engine is the same whether the car is being towed or being pushed but, if the car is to be towed, first read the previous Section on towing for details of where to connect the tow rope.

Proceed as follows:

1   Turn the ignition key to position **II** to switch on the ignition.

2   If the engine is cold, on models with a manual choke, pull out the choke control.

3   Depress the accelerator pedal.

4   Depress the clutch pedal and select third gear. Hold the clutch pedal down.

5   Tow or push the car, and slowly release the clutch pedal. The engine will turn over, and should start (don't worry about the car 'juddering' as the engine turns). If the car is being towed, as soon as the engine starts, depress the clutch pedal and *gently* brake the car so that you don't run into the tow vehicle.

STARTING USING JUMP LEADS

Note: *Starting using jump leads can be dangerous if done incorrectly. If you're uncertain about the following procedure, it's recommended that you ask someone suitably qualified to do the jump starting for you.*

Before attempting to use jump leads, there are a few important points to note.

First of all, use only proper jump leads which have been specifically designed for the job.

Make sure that the booster battery (the fully-charged battery) is a 12-volt type.

Position the two vehicles close enough together to connect the leads, but **do not** allow the vehicles to touch.

Turn off all electrical circuits.

DO NOT allow the ends of the two jump leads to touch at any time during the following procedure.

Proceed as follows.

1   Open the vehicle bonnets, and connect one end of one of the jump leads (usually the red one) to the positive (+) terminal of the flat battery, and the other end to the positive terminal of the booster battery.

2   Connect one end of the other jump lead

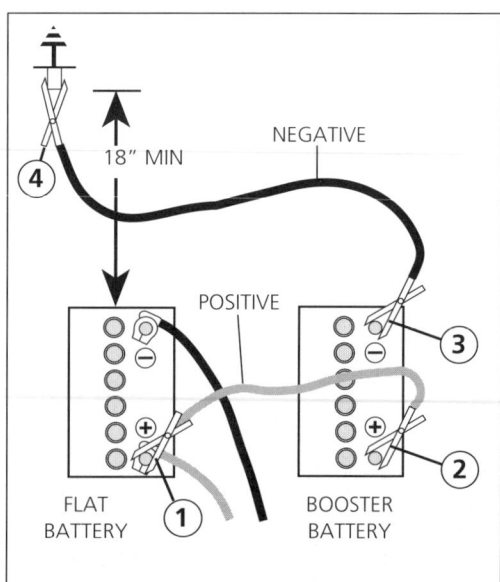

▲ *Jump start lead connections for negative earth vehicles – connect leads in order shown*

to the negative (-) terminal of the booster battery, and connect the other end to a suitable earth point on the car with the flat battery at least 50 cm (18 in) away from the battery (eg a clean, bare metal area on the engine block or body). Make sure that all the lead clips are secure.

3   Start the engine of the vehicle with the booster battery, let it run for a few minutes, then start the engine of the car with the flat battery in the normal way.

4   When the engine is running smoothly, disconnect the jump leads in exactly the reverse order to that in which they were connected.

What to carry in case of a breakdown

Carrying the following items may help to reduce the inconvenience and annoyance caused if you're unlucky enough to break down during a journey:

- Alternator drivebelt (refer to *'Servicing'* on page 106 for details of renewal)
- Spare fuses (refer to *'Bulb, fuse and relay renewal'* on page 135 for details of renewal)
- Set of principal light bulbs (refer to *'Bulb, fuse and relay renewal'* on page 127 for details of renewal)
- Battery jump leads (refer to *'Starting a car with a flat battery'* on page opposite for details of starting using jump leads)
- Tow-rope (refer to *'Towing'* on page 54 for details of towing procedure)
- Litre of engine oil (refer to *'Regular checks'* on page 83 for details of checking oil level)
- Torch

DRIVING SAFETY

In the UK, more people are injured or killed every year due to road accidents than through any other single cause.

Car manufacturers are paying more attention to passive safety when designing new cars, but drivers must ultimately take active responsibility of ensuring the safety of themselves, their passengers, and just as importantly, other road users.

There will always be accidents on the roads, but taking the time to read the following advice may help you to avoid such a mishap.

The following Sections aim to give advice which will help you to reduce the risk of becoming involved in an incident when driving, not necessarily by changing the way that you drive, but simply by making you aware of some of the potential risks which can easily be avoided. For more information and practical advice on road safety, it is well worth considering taking part in one of the various courses provided by organisations such as RoSPA and the Institute of Advanced Motorists.

BEFORE STARTING A JOURNEY

As a driver, before beginning a journey it's important to make sure that you're comfortable so that you can concentrate on driving without any unnecessary distractions. Spending a few moments carrying out a few simple checks and adjustments will help to make your journey more relaxing, safer and hopefully trouble-free.

Driver comfort

Before driving, make sure that you're comfortable, and that you can operate all the controls easily, particularly if someone else has recently driven the car.

● **Seat**
  Make sure that the seat is positioned a comfortable distance from the steering wheel and the pedals, so that you can operate all the controls comfortably

● **Seat back**
  Make sure that the seat back is adjusted to give your back plenty of support. Your back should rest against the seat, and there should be no need to lean forwards from the seat back. You should be able to reach the steering wheel comfortably, and you should be able to turn the wheel easily without stretching.

● **Head restraint**
  If a head restraint is fitted, it should be adjusted to support your head in the event of an accident. Your head should not rest against the restraint during normal driving and, as a rough guide, the restraint is correctly positioned when its top is in line with your eyes. Similarly, make sure that any passenger head restraints are adjusted correctly if passengers are being carried.

● **Mirrors**
  All the mirrors should be adjusted so that they give a clear view behind the car without the need to move your head unnecessarily.

● **Steering column**
  If the car is fitted with an adjustable steering column, adjust the column so that the steering wheel can be reached comfortably, and the wheel can be turned easily without stretching.

Checks

If you're planning a long journey, refer to *'Regular checks'* on page 81, and make sure that you carry out all the checks mentioned before setting off. Also make sure that you know where the spare wheel and the tools required for wheel changing are located (refer to *'Breakdowns'* on page 49).

DRIVING IN BAD WEATHER

When driving in bad weather, always be prepared. Bad weather should never take you by surprise. Listen to a weather forecast: you should always be aware of the possibility of poor conditions on your intended journey.

Driving in bad weather requires more concentration, and is more tiring than driving in good weather conditions. Never allow yourself to be distracted by talkative passengers or loud music in the car and, if you feel tired, stop at the next opportunity and take a break.

Always look well ahead, so that you're aware of the condition of the road surface, and any obstacles; **slow down if necessary**.

The following advice is intended to help you to drive more safely in various bad weather conditions. However, above all, use common sense.

Rain

● Use dipped headlights in poor visibility.
● Slow down if visibility is poor or if there is a lot of water on the road surface.
● Keep a safe distance from the vehicle in front (stopping distances are doubled on a wet road surface).

● Be particularly careful after a long period of dry weather. Under these conditions, rain can make the road surface very slippery.
● Don't use rear foglights unless visibility is **seriously** reduced (generally less than 100 metres/300 feet). Foglights can dazzle drivers following behind, especially in motorway spray.

Fog

● Slow down. Fog is deceptive, and you may be driving faster than you think. Fog can also be patchy, and the visibility may suddenly be reduced. Always drive at a speed which allows you to stop in the distance you can see ahead.
● Use dipped headlights. Using main-beam headlights will usually reduce the visibility even further, as the fog will scatter the light.
● Use foglights (where they are fitted) if it's genuinely foggy (generally, where visibility is reduced to less than 100 metres/300 feet), and not just misty.
● Keep a safe distance from the vehicle in front.
● Use your windscreen wipers to clear moisture from the windscreen.

Snow

● Don't start a journey if there is any possibility that conditions may prevent you from reaching your destination. Listen to a weather forecast before setting off.
● Before starting a journey, clear **all** snow from the windscreen, windows and mirrors. Don't just clear a small area big enough to see through.
● Slow down.
● Keep a safe distance from the vehicle in front (stopping distances can be trebled or even quadrupled on a snow-covered or icy surface).
● Drive smoothly and gently. Accelerate gently, steer gently and brake gently.
● Don't brake and steer at the same time – this may cause a skid (refer to 'Skid control' on page 64).
● When moving away from a standstill, or manoeuvring at junctions, etc, in a car with a manual gearbox use the highest possible gear that your car will accept to move away, and change up to a higher gear earlier than

usual. In a car with automatic transmission move the selector lever to prevent the transmission from using all the gears (usually position '2' – refer to *'Controls and equipment'* on pages 25 & 26). This will provide more grip and will help to reduce wheelspin.

● Use main roads and motorways where possible. Major roads are likely to have been 'gritted' and are usually cleared before minor roads.
● Use dipped headlights in poor visibility.
● Don't use rear foglights unless visibility is **seriously** reduced (generally less than 100 metres/300 feet) – foglights can dazzle drivers following behind.

Ice and frost

● Follow the advice given for driving in snow.
● Be prepared for 'black ice'. Although you can't see 'black ice', reduced road noise from the tyres will usually tell you it's there.

Severe winter weather

The best advice in severe winter weather is to stay at home but, if you must drive your car, in addition to following the advice given for driving on snow and ice, there are a few additional pieces of advice which you should bear in mind before setting out on a journey.

● Make sure that you tell someone where you're going, and tell them roughly what time you're expecting to arrive at your destination, and what route you're taking.
● Keep a can of de-icer fluid, a scraper, a set of jump leads and a tow rope in the car at all times.
● Carry plenty of warm clothes and blankets.
● Make sure that you have a full tank of petrol before starting your journey. This will allow you to keep the engine running, without fear of running out of petrol, to provide warmth through the car's heating system should you be delayed by conditions, or worse still stuck.
● Pack some pieces of old sacking or similar material, which you can place under the driving wheels to give better traction if you get stuck.
● Carry a shovel, in case you need to dig yourself or someone else out of trouble.

MOTORWAY DRIVING

Motorway driving requires a great deal of concentration and awareness. Motorways are generally busier than ordinary roads, and the traffic moves faster, so there is less time to react to any changes in road conditions ahead.

This Section covers the fundamental rules for safe motorway driving, and will help you to avoid many of the problems which occur on today's busy motorways. Remember that lack of common sense is probably the biggest cause of accidents on motorways.

Full rules and regulations for motorway driving can be found in 'The Highway Code'.

Joining and leaving a motorway

When you join a motorway, you will normally approach from a 'slip-road' on the left. You must give way to traffic already on the motorway. Watch for a safe gap in the traffic, and adjust your speed in the acceleration lane so that when you join the left-hand lane of the motorway you're already travelling at the same speed as the other traffic. Indicate before pulling onto the motorway. After joining the motorway, allow yourself some time to get used to the speed of the traffic before overtaking.

When you leave a motorway, indicate in plenty of time, and reduce your speed before you enter the slip-road – some slip-roads and link-roads between motorways have sharp bends which can only be taken safely by slowing down. Be very cautious immediately after leaving a motorway, as it can be difficult to judge speed after a long period of fast driving.

Rules for safe motorway driving

- **Concentrate and think ahead** – Always be prepared for the unexpected, and look well ahead. You should be aware of all the traffic in front and behind, not just the vehicle in front of you. If you're concentrating properly, nothing should take you by surprise on a motorway.
- **Always drive at a speed to suit the road conditions** – Don't break the speed limit, especially temporary speed limits which may apply to contra-flow systems or roadworks.
- **Slow down** in bad weather conditions. Driving too fast in fog and motorway spray is a major cause of motorway accidents.
- **Always keep a safe distance from the vehicle in front** – The closer you drive to the vehicle in front, the less chance you have of avoiding an accident if something happens ahead. Remember that you need more space in bad weather conditions.
- **Think** – If the vehicle ahead suddenly stops, do you have enough space to stop without hitting it?
- **Use your mirrors regularly** – It's just as important to be aware of what's happening behind you as it is in front.
- **Always signal your intentions clearly and in plenty of time** – Check your mirrors first, signal *before* you move, and change lanes smoothly and in plenty of time. Don't make any sudden moves which are likely to affect traffic approaching from behind.
- **Keep to the left** – The left-hand lane is *not* the slow lane, and you should drive in this lane whenever possible. You can stay in the middle lane when there are slower vehicles in the left-hand lane, but return to the left-hand lane when you've passed them. The right-hand lane is for overtaking only: it is *not* for driving at a constant high speed. If you use the right-hand lane, move back into the middle lane and then into the left-hand lane as soon as possible, but without cutting in.

- **Take note of direction signs** – You may need to change lanes to follow a certain route, in which case you should do so in plenty of time.
- **Stop as soon as practicable if you feel tired** – Driving on a motorway when feeling tired can be extremely dangerous. If necessary, wind down the window for fresh air. If you feel tired or drowsy, stop at the next service station, or turn off the motorway at the next junction and take a break before continuing your journey. You must not stop on the hard shoulder of a motorway other than in an emergency.
- **If you break down** – Refer to *'Breakdowns'* on page 48.

TOWING A TRAILER OR CARAVAN

When towing a trailer or a caravan, there are a few special points to bear in mind. There's more to towing a trailer or caravan than just simply bolting on a towbar and hitching up!

The law

Before using your car for towing, make sure that you're familiar with any special legislation which may apply, particularly if you're travelling abroad.

Make sure that you know the speed limits applicable for towing.

In some countries, including the United Kingdom, there is a legal requirement to have a separate warning light fitted in the car to show that the trailer/caravan direction indicator lights are working.

Always check on current regulations before you travel.

Before starting a journey

Before attempting to use your car for towing, there are one or two points which should be considered.

- **Make sure that your car can cope with the load which you're towing**
- **Don't exceed the maximum trailer or towbar weights for the car** – Refer to *'Dimensions and weights'* on page 14.
- **Engine** – Don't put unnecessary strain on your car's engine by trying to tow a very heavy load. Obviously, the smaller the engine, the less load the car can comfortably tow. Also bear in mind that the extra load on the engine when towing means that the engine's cooling system may no longer be adequate. Some manufacturers provide modified cooling system components for towing, such as larger radiators, etc, and if you intend to use your car for towing regularly, it may be worth enquiring about the availability and fitting of these components.
- **Suspension** – Towing puts extra strain on the suspension components, and also affects the handling of the car, since standard suspension components aren't usually designed to cope with towing. Heavy duty rear suspension components are available for most cars, and it's worth considering having these fitted if you intend to use your car for towing regularly. Special stabilisers are also available for fitting to most cars to reduce pitching and 'snaking' movements when towing.
- **Make sure that you can see behind the trailer/caravan using the car's mirrors** – Additional side mirrors with extended arms are available to fit most cars. The mirrors must be fitted with folding arms (so that they fold if hit), and should be adjusted to give a good view to the rear at all times. The mirrors can be removed when the car isn't being used for towing.
- **Make sure that the tyre pressures are correct** – Unless a light, unladen trailer is

being towed, the car tyres should be inflated to their 'full load' pressures. Also make sure that where applicable the trailer or caravan tyres are inflated to their recommended pressure.

- **Make sure that the headlights are set correctly** – Check the headlight aim with the trailer/caravan attached, and if necessary adjust the settings to avoid dazzling other drivers.
- **Make sure that the trailer/caravan lights work correctly** – Make sure that the tail lights, brake lights, direction indicator lights and, where applicable, the reversing lights and fog lights all work correctly. Note that the additional load on the direction indicator circuit may cause the lights to flash at a rate slower than the legal limit, and in this case you may need to fit a 'heavy duty' flasher unit.
- **Make sure that the trailer/caravan is correctly loaded** – Refer to the manufacturer's recommendations for details of loading. As a general rule, distribute the weight so that the heaviest items are as near as possible to the trailer/caravan axle. Secure all heavy items so that they can't move. To provide the best possible control and handling of the car when towing, the manufacturers recommend an optimum noseweight for the trailer/caravan when loaded (refer to *'Dimensions and weights'* on page 14 for details). The noseweight can be measured using a set of bathroom scales as follows:

    Place a stout piece of wood between the trailer/caravan tow hitch cup and the scales platform, and read off the weight with the trailer/caravan level. If necessary, redistribute the load, to arrive as close as possible to the recommended noseweight. **Do not** exceed the maximum recommended noseweight.

Driving tips

Towing a trailer or caravan will obviously affect the handling of the car, and the following advice should be followed when towing.

- **If possible, avoid driving with an unladen car and a loaded trailer/caravan** – The uneven weight distribution will tend to make the car unstable. If this is unavoidable, drive slowly to allow for the instability.

● **Always drive at a safe speed** – The stability of the car and the trailer or caravan decreases as the speed increases. Always drive at a speed which suits the road and weather conditions, and always reduce speed in bad weather and high winds – especially when driving downhill. If the trailer or caravan shows any sign of 'snaking', reduce speed immediately – never try to stop 'snaking' by accelerating.

● **Always brake in good time** – If you're towing a trailer or caravan which has brakes, apply the brakes gently at first, then brake firmly. This will help to prevent the trailer wheels from locking. In cars with a manual gearbox, change into a lower gear before going down a steep hill so that the engine can act as a brake, and similarly, in cars with automatic transmission, move the selector lever to position '2', or '1' in the case of very steep hills.

● **Don't change to a lower gear unnecessarily** – Unless the engine is labouring, stay in as high a gear as possible to keep the engine revs as low as possible. This will help to avoid the engine overheating.

ALCOHOL AND DRIVING

By far the best advice on drinking and driving is **DON'T**.

You may feel fine, but it's a proven fact that even one small alcoholic drink will impair your driving to some extent.

Drinking alcohol has the following effects:

● **Reduces co-ordination**
● **Increases reaction time**
● **Impairs judgement of speed, distance and risk**
● **Encourages a false sense of confidence**

The risk of an accident increases sharply after drinking alcohol, and approximately one third of the total number of people killed in road accidents each year have blood alcohol levels above the legal limit for driving. Remember that you're putting other people as well as yourself at risk if you drive after drinking.

The legal limit for blood alcohol level in the UK is 80 milligrams per 100 millilitres. This doesn't correspond to any particular quantity of drink, as the amount of drink required to reach this level varies from person to person. The driving of many people who feel perfectly sober is seriously affected well below the legal limit.

The penalties for driving over the legal limit are severe, and can mean losing your driving licence, a heavy fine, or imprisonment. It's also important to realise that the laws abroad can be far more severe than in the UK, and some countries have a total ban on driving after drinking alcohol.

● **The safest course of action is not to drink and drive.**

SKID CONTROL

If you drive sensibly with due regard for the road conditions, you should never find yourself in a situation where your car is skidding. The following advice will help you to avoid situations which may cause a car to skid, and explains how to regain control of your car quickly should the need arise.

A skid is caused by one or a combination of the following:

- **Excessive speed in relation to the road conditions.**
- **Harsh or excessive acceleration.**
- **Sudden or excessive braking.**
- **Coarse or excessive steering.**

The most common basic cause of skidding is rough handling of the car's controls. Therefore the key to safe driving and preventing a skid is smooth, gentle handling of the car and its controls. Try to apply smooth pressure to the brake pedal, accelerator and steering wheel rather than just suddenly moving them a certain distance.

When a car is skidding, the following things happen:

- **The car is out of control. You can take action to regain control, but while skidding, the car can't be fully controlled.**
- **A car with the front wheels skidding can't be steered.**
- **A car with the rear wheels skidding is likely to spin round if any steering lock is applied.**
- **A car with all four wheels skidding will continue in a straight line in the direction it was travelling when the skid started, regardless of which way the car is pointing.**

Prevention is better than cure. When approaching a corner, reduce speed early by smooth progressive braking **in a straight line**. In a car with a manual gearbox, brake to an appropriate speed and select the correct gear for the corner. Turn into the corner early and smoothly, gently increasing the steering lock if necessary, and corner with your foot gently on the accelerator, so that the engine just 'pulls' the car through the corner, maintaining a constant speed. As you leave the corner, accelerate gently and smoothly.

If you follow the above advice, you should avoid finding yourself in a situation where your car is skidding. However, should you be unfortunate enough to experience skidding, the basic rule for skid control is to **remove the cause** of the skid. If you cause a skid by accelerating, braking or steering, **ease off**, and, when the skid stops, re-apply the accelerator, brake or steering control, but this time more gently and smoothly. If the front wheels are skidding, to regain steering control quickly, steer 'into the skid'. This is often misunderstood, and it basically means look in the direction you want to be facing, and steer in that direction – and be prepared to reduce the steering quickly if necessary when the wheels regain their grip.

Several organisations offer courses in skid control, using specially-equipped cars under carefully-controlled conditions. One of these courses could prove to be a very worthwhile investment, possibly helping you to avoid an accident.

ADVICE TO WOMEN DRIVERS

Unfortunately, it's a fact that women are more likely than men to be the subject of unwelcome attention when driving alone.

There's no reason to think that driving alone spells trouble, but it's a good idea to be aware of certain precautions which will help to avoid finding yourself in an unpleasant situation.

This Section aims to give advice which will help to reduce any risks when driving alone, and will help you to deal with the situation, should you be unfortunate enough to find yourself the subject of unwelcome attention.

Several organisations (some local police authorities, AA, etc) run courses especially for women drivers, in which subjects such as basic car maintenance and self defence, etc, are covered.

Sensible precautions
● The car

Make sure that you're familiar with your car.

Make sure that your car is regularly serviced. Refer to 'Regular checks' on page 81, and make sure that all the checks described are carried out regularly. This will minimise the possibility of a breakdown.

Make sure that you know how to change a wheel, and make sure that the car jack is in good condition in case you have a puncture - refer to 'Breakdowns' on page 49.

● The journey

If you're planning to undertake a particularly long or unfamiliar journey, the following advice may be helpful.

If you're travelling on an unfamiliar route, make a few notes before you set off, reminding yourself of the road numbers, where to turn, which junctions to use on motorways, etc. Try to stick to main roads where possible.Always carry a map.

Make sure that you have enough petrol for the journey. If you need to fill up during the journey, make sure that you do so in plenty of time, before the fuel level gets too low, and before petrol stations close if travelling at night.

If you think that your family or friends may be worried, you may want to 'phone someone at your destination before setting off to tell them that you're leaving, and tell them what time you expect to arrive. Similarly, when you arrive, you may want to 'phone someone at your starting point to confirm that you've arrived safely.

● Driving in traffic

Avoid attracting unnecessary attention.

Avoid 'jokey' stickers in car windows, which may encourage unwelcome attention.

Avoid eye contact with 'undesirables'.

If you're being followed, pull over and slow down, but **don't** stop. **Don't** drive to your home, but find a busy and well lit place. If the person following persists, blow your horn and flash your lights to attract attention.

If you're stopped by traffic or another vehicle, lock the doors and close the windows. **Don't** ram the other vehicle - damage to your car might prevent your escape.

● Leaving your car

Don't leave any clues that the car is being driven by a woman; make sure that any 'feminine' items are hidden from sight before leaving your car.

Refer to *'Car crime prevention'* on page 75 and take note of the advice given.

Try to park in a well-lit, preferably busy area.

If you park in a car park, try to park close to an exit, or close to the attendant's station. **Always** reverse into a parking space, so that you can drive away quickly if necessary after returning to your car.

Take note of any 'landmarks' so that you can find your car quickly when you return. **Always** lock your car.

When you return to your car, walk with a group of people, if possible. Have the keys ready so that you don't have to spend unnecessary time outside your car searching for them, which may attract attention.

Before getting into your car, briefly check for forced entry, and look into the car for any suspicious signs. **Don't** get into the car if you notice anything suspicious.

● Breakdowns

Remember that you're more likely to be injured in an accident than a personal attack.

Refer to *'Breakdowns'* on page 47 and take note of the advice given but, in addition, bear in mind the following advice.

Always walk to 'phone for assistance, **don't** accept a lift. When you 'phone for assistance, mention that you're an unaccompanied woman.

If someone stops while you're 'phoning for help, give the operator details of the other vehicle's registration number and a brief description of the car and driver. If the driver approaches you, tell them that you have passed on his/her details to the police. If the driver's intentions are honourable, your reaction will be understood.

If you decide to stay with, but outside the car, leave the nearside door unlocked and slightly open, so that you can get inside quickly to lock yourself in if necessary.

If you decide to stay inside the car, sit in the passenger seat and lock the door - this will give the appearance that you're accompanied.

When help arrives, ask for some form of identification (even from a policeman) before giving any of your own details.

If someone offers assistance, tell them the police have been informed and are arranging recovery. If you have not yet contacted the police, consider asking the person to do so on your behalf, **but** if you're uncertain about the person's intentions, tell them that the police are aware and ask them to call the police again for you.

● Accidents

Keep calm.

Refer to 'Accidents and emergencies' on page 39 and take note of the advice given but, in addition, bear in mind the following advice.

If you're bullied or shouted at, lock yourself in your car (if it's safe to do so), and wind the windows up. Communicate through a small gap at the top of the window.

If things get out of hand, refuse to talk to anyone except the police.

Self defence

This is a last resort and should not be used unless all else has failed.

There is no substitute for taking a properly supervised course in self-defence, but we aim to outline a few of the basic moves below.

If you are approached, act in a composed manner and use a soft tone.

In the event of an attack, stay calm - you can still safeguard yourself.

- **If you're held by one arm** - Use your free hand to grasp the attacker's thumb and twist the thumb sharply back towards the wrist.
- **If you're facing your attacker** - If your shoulders are held, thrust your hands/arms upwards and diagonally outwards to dislodge the attacker's grip.

  Use your knees or the blade of your hand to chop your attacker's groin, your feet to kick the shins, and your fingers to poke the eyes.
- **If you're attacked from behind** - Quickly lean forward, twisting your head sideways into the attacker to keep your airway (for breathing) clear. Often this will cause the attacker to lose balance.
- **If you've been pulled backwards already** - Turn your head as above, and chop hard with the blade of your hand or a clenched fist to the groin, or an elbow in the stomach.
- **If you've been forced to the floor** - Use your feet and legs to kick against the attacker's shins whilst swivelling your body to keep the attacker at bay.

Be ready to run as soon as the attacker's grip is released, and SCREAM!

CHILD SAFETY

Every day, thousands of parents strap their offspring into child car seats, confident that they have done everything possible to protect their loved ones. However, a recent survey has shown that many young children are travelling in car seats which have been incorrectly fitted or are being incorrectly used, making them potentially dangerous should they be involved in an accident. The following advice will help you to make sure that you take every possible precaution to ensure the safety of your children when driving.

How to avoid unnecessary risks

- **Never allow young children to travel in a car unrestrained,** even for the shortest of journeys.
- **Never carry a child on an adult's lap or in an adult's arms.** Although you may feel that your baby is safer in your arms, this is not the case. An adult holding a child is far more likely to cause injury to the child than to give protection in the event of an accident.

- **Never rely solely on an adult seat belt to restrain a child,** and never sit a child on a cushion to enable a seat belt to fit properly.
- **Always strap young children into a properly designed child car seat** when carrying them in a car. The cost of a good car seat is a very small price to pay to save your child's life.

Choosing a child car seat

- It's vitally important to choose the correct type of car seat for your child. A wide range of child car seats is available, and it's worth spending some time looking at the various seats on the market before deciding on the most suitable seat for your particular requirements.
- Although age ranges are often given by the manufacturers, these should be taken as a rough guide. It's the weight of the child which is important; for example, a smaller than average baby could use a baby car seat for longer than a heavy baby of the same age.
- **Never** buy a secondhand child seat. This may sound like a ploy from the manufacturers to sell more seats, but all too often a secondhand seat is sold without the instructions, and sometimes there are parts missing. This often leads to secondhand seats being incorrectly fitted. A secondhand seat may have been damaged or weakened through carelessness or misuse, without necessarily showing any visible signs until it's put to the test in an accident!
- Your baby's first contact with the outside world is often on that first ride home from hospital, so make sure that you're prepared, and buy a car seat before your baby is born.
- Some seats are designed for babies from birth to a weight of around 22 pounds (10 kg), which for most babies is around nine months old. Usually, these first car seats are light and easy to transport through the use of a handle. This means that a sleeping baby can be carried from the car into the house without waking. Some baby seats come complete with a built-in headrest, which is very important for the early weeks when a baby's head needs to be supported.

Certain seats can be fitted so that the baby faces the back of the car (ie rearward facing seats). This may be considered to improve safety, as the baby is supported across the back rather than purely by the harness, if the car is involved in a frontal impact. Combination type seats can then be used forward facing when the child reaches approximately nine months old.

Other types of seat may be designed to accommodate children up to around 40 pounds (18 kg) or approximately four years old.

Alternatively, some seats use the car's seat belts to hold both the child seat and the child, and these seats are obviously easier to fit. Make sure that this type of seat is fitted with a seat-belt lock, so that the seat belt cannot be pulled out of place or slackened. Another advantage of these seats is that they can easily be transferred from car to car.

- Try to choose a seat which has an easily adjustable harness, as this makes it easier to ensure that the harness fits the child securely for each trip. If the harness can be quickly loosened, it makes getting a struggling child in and out a little easier.

Using a child car seat

- Firstly, make absolutely sure that the seat is properly fitted in accordance with the manufacturer's instructions. If you're unsure about any of the fitting procedure, contact the manufacturer for advice.
- To hold a child securely, the harness must be reasonably tight. There should be just enough room to slide your flat hand under the strap. Children may be wearing bulky clothes one day, and thin clothes the next, so it is vital that the harness is adjusted before each journey to ensure a correct fit.
- If the seat is fitted using an adult's inertia reel seat belt, make sure that the seat is held firmly in position, and that the seat belt is securely locked to prevent it from loosening. Also, make sure that the seat belt buckle is not resting on the frame of the child seat. This is because the buckles are not designed to withstand the impact of a heavy child seat and, in an accident, could break open.

DRIVING ABROAD

Driving abroad can be very different to driving in the UK. Besides the obvious differences, like climate and driving on the right-hand side of the road, various unfamiliar laws may apply, and it's advisable to prepare yourself and your car as far as possible before travelling.

This Section provides you with a guide which will help you to prepare for driving abroad, and will help you to avoid some of the pitfalls waiting for the unwary.

It may be worthwhile considering hiring a car abroad, rather than driving your own car. In this case, make sure that the insurance cover arranged suits your requirements, and make sure that a damage waiver is included (otherwise you will have to pay for any damage to the hire car).

Insurance
● Motoring
Make sure that you have adequate insurance cover for your car and your luggage.

Check on the legal requirements for insurance in the country you're visiting, and always inform your insurance company that you're taking your car abroad – they will be able to advise you of any special requirements.

Most car insurance policies automatically give the minimum legally-required cover for driving in EC countries, but if you require the same level of cover as you have in the UK, you will normally need to obtain a 'Green Card' (an internationally-recognised certificate of insurance) from your insurance company.

● Medical
It's always advisable to take out medical insurance for the car occupants. Not all countries have a free emergency medical service, and you could find yourself with a large unexpected bill in the event of yourself or one of your passengers being taken ill, or being involved in an accident (in some countries you may even have to pay for an ambulance).

NHS form E111 (available by filling in a form at a Post Office) entitles you to receive

AUSTIN/MG MAESTRO & MONTEGO

the same health care that residents receive in EC countries, but this is by no means comprehensive, and additional insurance cover is usually advisable.

● Breakdown

Recovery and breakdown costs can be far higher abroad than in the UK.

Most of the national motoring organisations (such as the AA and RAC) will be able to provide insurance cover which could save you a lot of inconvenience and expense should you be unfortunate enough to break down.

Documents

Always carry your passport, driving licence, car registration document, and insurance certificate (including 'Green Card' and medical insurance, where applicable).

Make sure that all the documents are valid, and that the car's road fund licence (tax) and MOT don't run out while you're abroad.

Before travelling, check with the authorities in the country you're visiting, in case any special documents or permits are required. You may need a visa to visit some countries, and an International Driving Permit (available from the AA or RAC) is sometimes required.

Driving laws

Before driving abroad, make sure that you're familiar with the driving laws in the country you're visiting, as there may be some laws which don't apply in the UK, and the penalties for breaking the law may be severe.

Fit a 'GB' plate to the back of your car, and make sure that it's displayed all the time you're abroad.

In some countries, you're legally required to carry certain items of safety equipment. These can include a first-aid kit, a warning triangle, a fire extinguisher, a set of spare bulbs/fuses, etc.

Remember that if you're visiting a country where you have to drive on the right-hand side of the road, you'll need to fit headlight beam deflectors or shields to avoid dazzling other drivers, or alternatively, it may be possible to have the headlight beams adjusted.

Make sure that you're familiar with the speed limits, and note that in some countries there is an absolute ban on driving after drinking *any* alcohol.

Servicing

Service your car before setting off on your trip, to reduce the possibility of any unexpected breakdowns.

Pay particular attention to the condition of the battery, windscreen wipers and tyres (including the spare), noting that the tyre pressures will probably have to be increased from their normal setting if the car is to be fully loaded. If you're travelling to a cold country, check the condition of the cooling system and the strength of the antifreeze.

Before setting off on your journey, refer to *'Regular checks'* on page 81, and carry out all the checks described. Check that the jack and wheel brace are in place in the car (refer to *'Breakdowns'* on page 49), and that the jack works properly.

Spares

In addition to the items which must be carried by law in the country you're visiting, it's a good idea to carry a few spares which may be difficult to obtain should you need them abroad (one of the national motoring organisations should be able to advise you). For example, you may want to carry clutch and throttle cable repair kits, as right-hand-drive components can be difficult to find outside the UK.

If your car uses a special oil, it's a good idea to take a pack with you.

It's also a good idea to carry a tow rope and a set of jump leads to help you out in case of a breakdown.

Fuel

The type and quality of petrol available varies from country to country, and it's a good idea to check on the availability of the correct petrol type before travelling (again, one of the national motoring organisations will be able to advise you). This is especially important if your car has a catalytic converter, as in this case you must only use unleaded petrol.

Find out what petrol pump markings to look for to give you the correct type and grade of petrol for your car.

Security

A foreign car packed with luggage is an inviting prospect for criminals, so refer to *'Car crime prevention'* on page 75, and don't take any chances.

Route planning

It's always a good idea to plan your approximate route before travelling. A vast number of maps and guides are available, or the AA and RAC can provide you with directions to your destination for a modest charge.

Bear in mind that in some countries, you'll have to pay tolls to use certain roads, and this can add unexpectedly to the cost of travelling.

What to carry when driving abroad

The following list provides a guide to the items which it is compulsory to use or carry, or it is strongly recommended that you carry in your car when driving in various European countries.

In the following table '**C**' indicates '**Compulsory**' and '**R**' indicates '**Recommended**'.

COUNTRY	HEADLAMP DEFLECTORS	GB STICKER	SEAT BELTS	WARNING TRIANGLE	FIRE EXTINGUISHER	FIRST AID KIT	SPARE BULBS
Austria	C	C	C	C	-	C	-
Belgium	C	C	C	C	C	-	-
Bulgaria	C	C	C	C	C	C	-
Czechoslovakia	C	C	C	C	-	C	C
Denmark	C	C	C	R	-	-	-
Eire	-	C	C	-	-	-	-
Finland	C	C	C	C	-	-	-
France	C	C	C	C	-	-	R
Germany	C	C	C	C	C	C	R
Greece	C	C	C	C	C	C	-
Holland	C	C	C	C	-	-	C
Hungary	C	C	C	C	-	-	C
Italy	C	C	C	C	-	-	R
Luxembourg	C	C	C	R	-	-	-
Norway	C	C	C	R	-	-	R
Poland	C	C	C	C	-	-	R
Portugal	C	C	C	C	C	-	-
Spain	C	C	C	C	-	-	C
Sweden	C	C	C	R	-	-	-
Switzerland	C	C	C	C	-	-	-
United Kingdom	-	-	C	-	-	-	-
Yugoslavia	C	C	C	C	-	C	C

AUSTIN/MG MAESTRO & MONTEGO

Owning a car is always going to involve some expense, and the running costs can generally be divided into two main areas. The first concerns fixed costs which cannot be avoided, such as car tax and insurance (although obviously you can shop around for the best insurance quote). However, for most owners savings can be made in the second main area of expense, which covers fuel costs and servicing/maintenance bills.

It's surprisingly easy to reduce the amount of fuel used, simply by adapting your driving style to suit the prevailing conditions, and avoiding certain driving habits which tend to increase fuel consumption unnecessarily.

A large proportion of the average servicing bill is made up of garage labour time, money which can be saved by carrying out the work yourself. Refer to *'Servicing'* on page 89 for easy-to-follow instructions on how to carry out most servicing work. It's also worth bearing in mind that maintenance costs can be cut significantly by reducing unnecessary wear-and-tear on the car.

The following advice deals with the most significant causes of high fuel consumption and unnecessary wear-and-tear, and explains how to save money and reduce environmental pollution during everyday driving.

● **Don't warm the engine up with the car standing still** – Engine wear and pollution is at its highest when the engine is warming up, and the engine takes a long time to warm up when running at idle speed with the car stationary. To avoid excessive wear and pollution, drive off as soon as the engine starts, and don't use more 'revs' than necessary.

● **On cars with a manual choke, don't use the choke any longer than necessary** – Push the choke control fully in as soon as the engine will run smoothly without it. When the engine is running with the choke applied, extra fuel is being used and so the fuel consumption is increased, and more pollution is produced.

● **Avoid sudden full throttle acceleration** – Sudden acceleration increases fuel consumption, engine wear and pollution.

● **Don't drive at high engine speeds** – Minimum fuel consumption and pollution is achieved at low engine speeds and in the highest possible gear. Lower engine speed also means less noise and engine wear. For maximum economy, stay in as high a gear as possible for as long as possible, without making the engine labour.

● **Don't always drive at maximum speed** – Fuel consumption, pollution and noise increase rapidly at high speeds. A small reduction in speed (particularly during motorway driving) can significantly lower fuel consumption and pollution.

● **Look well ahead, and drive as smoothly as possible** – By looking well ahead you will be able to react to any change in road conditions in plenty of time, allowing you to brake and accelerate smoothly. Unnecessary or harsh acceleration and braking increases fuel consumption and pollution.

● **Where possible, avoid dense, slow-moving traffic** – In these conditions, more frequent braking, acceleration and gear changing are required, which increase fuel consumption and pollution.

● **Stop the engine during traffic hold-ups** – Obviously if your engine has stopped it doesn't use any fuel and there is no pollution.

ECONOMICAL DRIVING

● **Check the tyre pressures regularly** – Low tyre pressures increase the rolling resistance of the car, and therefore increase fuel consumption, as well as increasing tyre wear and causing handling problems.

● **Don't carry unnecessary luggage** – Weight has a significant effect on fuel consumption, especially in dense traffic where frequent acceleration is required.

● **Don't leave a roof rack fitted when not in use** – The extra air resistance increases fuel consumption.

● **Switch off any unnecessary electrical circuits as soon as possible** – Heated rear windows, foglights, heater blowers, etc, consume a considerable amount of electrical power. The engine must work harder to provide this power by driving the alternator, and the fuel consumption is therefore increased.

● **Check the fuel consumption regularly** – By doing this you will be able to notice any significant increase in fuel consumption, and any problem which may be causing it can be investigated before it develops into anything more serious.

● **Ensure that your car is serviced regularly** – This will ensure that the car operates as efficiently as possible, reducing fuel consumption and pollution.

● **If your car engine is suitable, use unleaded petrol** – This will reduce pollution. If in any doubt as to whether your car's engine is suitable for use with unleaded petrol, seek advice from the car's manufacturer, or from a recognised dealer.

Crime against cars is a serious problem, and many owners suffer the consequences of theft or damage every year. The likelihood of your car becoming a target for criminals can be reduced by making it more difficult for your car to be broken into or stolen, and the following advice should prove helpful. Some of the points may seem obvious, but the majority of cars are broken into or stolen in a very short space of time with little force or effort required.

- **Always remove the ignition key** – even in a garage or driveway at home. Always make sure that the steering column lock is engaged before removing the key.
- **Always lock your car** – even in a garage or driveway at home. Where fitted, ensure that 'deadlocks' are engaged before leaving the car and, where applicable, don't forget to lock the fuel filler cover. Make sure that all the windows and the sunroof or folding roof, where applicable, are properly shut. If the car is in a garage, lock the garage. If your car is stolen or broken into while it's unlocked, your insurance company may not pay for the full value of loss or damage.

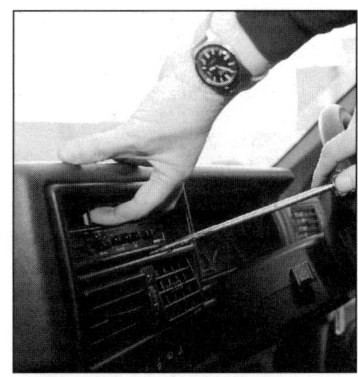

- **Never leave valuable items on display** – Even if you're only leaving your car for a few minutes, move anything that might be attractive to a thief (even a coat or briefcase) out of sight, and preferably take it with you or lock it safely in the boot. Don't leave valuable items (especially credit cards) in the glovebox. Don't leave your vehicle documents in the car (registration document, MOT certificate, insurance certificate, etc), as they could help a thief to sell it.
- **Park in a visible and (preferably) busy area** – This will deter thieves as they run a greater risk of being caught. If you're parking your car at night, try to leave it in a well-lit area.
- **Put your radio aerial down (where applicable) when parking** – Radio aerials can prove an attractive target for vandals.
- **Protect any in-car entertainment equipment** – The latest security-coded equipment won't work if someone tampers with it and disconnects it from the battery. Some equipment is specially designed so that you can remove the control panel or the entire unit, and take it with you when you leave the car.
- **Fit lockable wheel nuts (or bolts, as applicable)** – if your car is fitted with expensive alloy wheels. Alloy wheels are a favourite target for thieves.
- **Have your car windows etched with the registration number** – This will help to trace your car if it's stolen and the thieves try to change its identity. Other glass components such as sunroofs and headlamps can also be etched if desired. Many garages and specialists can provide this service, and it's possible to buy DIY glass etching kits from motor accessory shops if you prefer to tackle the job yourself.
- **Fit a vehicle immobiliser device** – Many different types are available, but the most common types consist of a substantial metal bar which can be locked in place between the steering wheel and the pedals to prevent the car from being driven.
- **Have an alarm fitted** – Many different types are available, and some are expensive, but they will deter thieves. Some alarms have built-in immobiliser devices to prevent the car from being driven. If you have an alarm fitted, remember to switch it on even if you're only leaving the car for a few minutes.

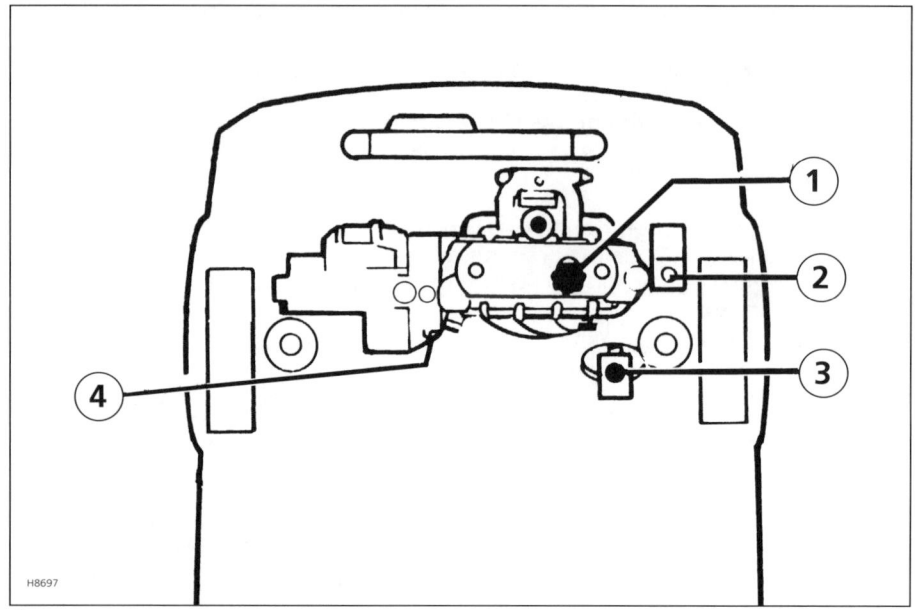

▲ 1.3 litre models & Maestro 1.6 litre models with 'R-series' engine

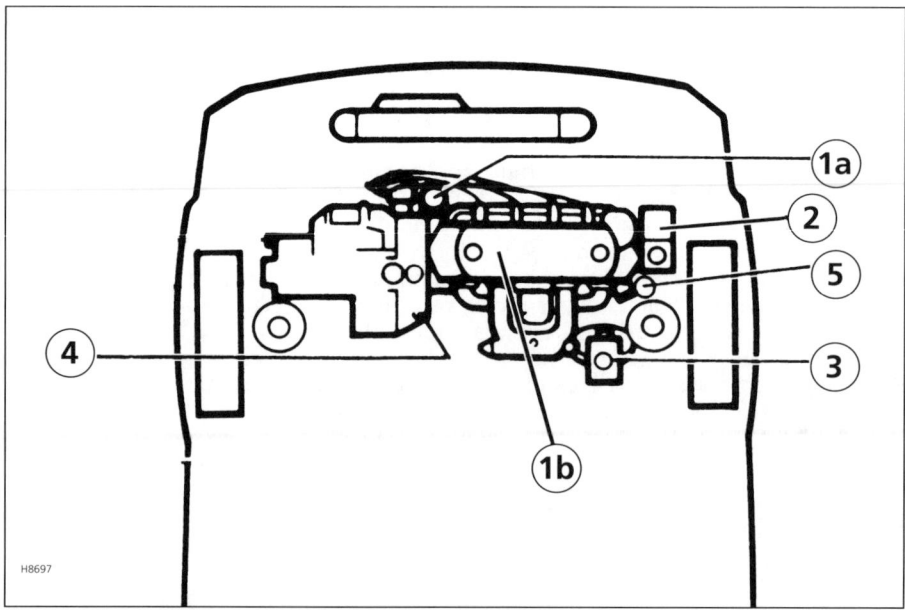

▲ Maestro 1.6 litre 'S-series' models, & Montego 1.6 litre models
  1a Models up to 1988 (approx)
  1b Models from 1988 (approx)

SERVICE SPECIFICATIONS

Recommended lubricants and fluids

Component/system	Lubricant/fluid type and specification	Duckhams recommendation
ENGINE (1)	Multigrade engine oil, viscosity SAE 15W/50, 10W/40 or 10W/30 (pre-August 1983) or 10W/40 (August 1983 on)	Duckhams QXR, Hypergrade, or 10W/40 Motor Oil
COOLING SYSTEM (2)	Mixture ratio two-thirds clean water to one-third ethylene glycol-based antifreeze	Duckhams Universal Antifreeze and Summer Coolant
BRAKING SYSTEM (3)	Hydraulic fluid meeting specifications FMVSS 116, DOT 4, or SAE J1703	Duckhams Universal Brake and Clutch Fluid
AUTOMATIC TRANSMISSION (4)	Automatic Transmission Fluid (ATF) to specification Dexron II D	Duckhams Uni-Matic or D-Matic
POWER-ASSISTED STEERING (5)	Automatic Transmission Fluid (ATF) to specification Dexron II D	Duckhams Uni-Matic or D-Matic

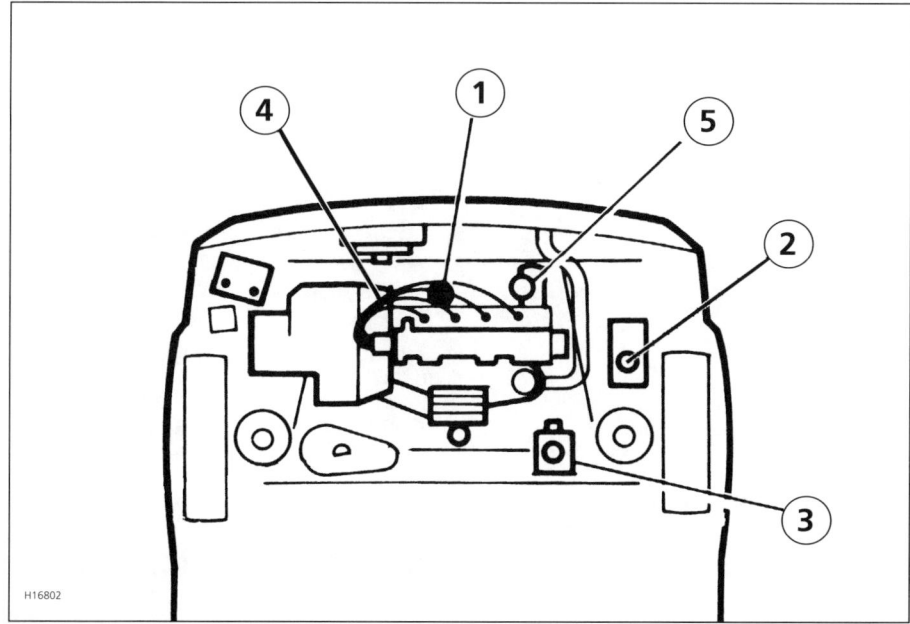

H16802

▲ *All 2.0 litre models*

AUSTIN/MG MAESTRO & MONTEGO

Lubricant and fluid capacities [Litres (Pints)]

Fuel tank capacity

1983 Maestro models	**53** (11.66 gallons)
All other models (approximate)	**50** (11.00 gallons)

Engine oil

Quantity of oil required to bring level on dipstick from 'MIN' to 'MAX' mark	**0.5** (0.88)

Engine oil capacity (for oil and filter change)

1.3 litre models	**2.7** (4.75)
Maestro 1.6 litre models with 'R-series' engine	**3.1** (5.50)
All other 1.6 litre models	**3.8** (6.75)
2.0 litre models	**4.8** (8.50)

Coolant capacity (for coolant change)

1.3 litre models	**6.7** (11.75)
Maestro 1.6 litre models with 'R-series' engine	**8.2** (14.50)
All other 1.6 litre models, 2.0 litre models	**8.5** (15.00)

Spark plugs

Note: *The following is intended as a guide only. Recommendations have changed, as have the spark plug types themselves, during the car's life. As the information given is based on the latest available, it may not correspond with that given elsewhere; if in doubt, consult a Rover dealer for specific information.*

Type	Unipart	Champion
1.3 litre models	**GSP 4362** or (1990-on models with catalytic converters) **GSP 3372**	**RN9YCC** or **RN9YC**
Maestro 1.6 litre models (except MG) with 'R-series' engine	**GSP 4362**	**RN9YCC** or **RN9YC**
MG Maestro 1600 with 'R-series' engine	**GSP 4452**	**RN7YCC** or **RN7YC**
Maestro 1.6 litre models (except MG) with 'S-series' engine, Montego 1.6 litre models up to 1988	**GSP 4562**	**RC9YCC** or **RC9YC**
MG Maestro 1600 with 'S-series' engine	**GSP 4552**	**RC7YCC** or **RC7YC**

Spark plugs (continued)

1989 Montego 1.6 litre models	**GSP 4562** or **GSP 3562**	**RC9YCC** or **RC9YC**
1990-on Montego 1.6 litre models	**GSP 4662** or **GSP 3662**	**RC9YCC** or **RC9YC**
Maestro 2.0 litre models	**GSP 4452**	**RN7YCC** or **RN7YC**
Montego 2.0 litre carburettor-engine models up to late 1988	**GSP 4362**	**RN9YCC** or **RN9YC**
1989-on Montego 2.0 litre carburettor-engine models	**GSP 4352** or **GSP 3362**	**RN7YCC** or **RN7YC**
Montego 2.0 litre fuel injection-engine models	**GSP 4452** or (1990-on models only) **GSP 3462**	**RN7YCC** or **RN7YC**
MG Montego Turbo up to 1989	**GSP 4352** or (1990-on models only) **GSP 3362**	**RN7YCC** or **RN7YC**
1990-on MG Montego Turbo	**GSP 4452** or **GSP 3462**	**RN7YCC** or **RN7YC**

Electrode gap [mm (in)]

Unipart plugs:

All models up to late 1988	**1.00** (0.039)
All models, late 1988-on	**0.85** (0.033)
Champion plugs	**0.80** (0.032)

Tyre pressures - cold [bars (lbf/in^2)]

Note: *The following is intended as a guide only. Manufacturers frequently change tyre pressure recommendations, and it is suggested that a Rover dealer is consulted for latest recommendations.*

Maestro models	Front	Rear
145 SR 13 or 145 R 13 74S tyres	**2.1** (30)	**2.2** (32)
155 SR 13 or 155 R 13 78S, 165 SR 13 or 165 R 13 82S tyres	**1.8** (26)	**1.9** (28)
175/65 SR 14, 175/65 HR 14 or 175/65 R 14 88H tyres	**1.9** (28)	**1.8** (26)
185/55 HR 15 tyres – 1989 models *	**2.2** (32)	**2.2** (32)
185/55 HR 15 tyres – 1990-on models *	**2.1** (30)	**2.1** (30)

* *If carrying more than four passengers and luggage, increase front and rear tyre pressures by **0.1** (2).*
   *MG EFi/2.0i models only – if travelling at more than 100 mph (where permitted), increase front and rear tyre pressures by **0.3** (4).*
   *MG Turbo models only – if travelling at more than 100 mph (where permitted), increase front and rear tyre pressures by **0.5** (7).*

Montego models	Front	Rear
1984 models	**1.8** (26)	**1.9** (28)
Later 1.3 & 1.6 litre Saloon models *	**1.8** (26)	**1.9** (28)
Later 2.0 litre Saloon models *	**1.9** (28)	**1.9** (28)
Later Estate models *	**1.9** (28)	**2.1** (31)

* *If carrying more than four passengers and luggage, increase front tyre pressure by **0.2** (3), Saloon rear tyre pressure by **0.1** (2), Estate rear tyre pressure by **0.5** (7).*
   *MG models up to 1987 – if travelling at more than 100 mph (where permitted), increase front and rear tyre pressures by **0.3** (4).*
   *MG models 1988-on – if travelling at more than 100 mph (where permitted), increase front and rear tyre pressures by **0.5** (7).*

To ensure that your car is reliable and safe to drive, there are one or two essential checks which are so simple that they're often ignored. These checks only take a few minutes, and could save you a lot of inconvenience and expense. It's a good idea to carry out these checks once a week, and certainly before you start off on a long journey. This Section explains how to carry out the checks, and what to do if things aren't quite as they should be.

Whenever you're carrying out checks or servicing jobs, safety must always be the first consideration, and you should bear in mind the advice given in the *'Safety first!'* notes on page 96 before proceeding.

CHECKS

The following checks should be carried out regularly. The checks are explained in more detail in the following pages.

- *Check the oil level*
- *Check the coolant level*
- *Check the brake fluid level*
- *Check the tyres*
- *Check the washer fluid level*
- *Check the battery electrolyte level (where applicable)*
- *Check the wipers and washers*
- *Check the lights and horn*
- *Check for fluid leaks*

ITEMS REQUIRED WHEN CARRYING OUT CHECKS

When carrying out the regular checks, it's a good idea to have the following items close-to-hand. You won't always need all the items, but it's as well to have them available just in case.

- *Small quantity of clean rag* ● *1.0 litre pack of engine oil (of the correct type – refer to **'Service specifications'** on page 77)* ● *Small quantity of coolant solution (made up from approximately two-thirds clean water and one-third antifreeze)* ● *Small container of brake fluid (must be an airtight container)* ● *Tyre pressure gauge and foot pump* ● *Small screwdriver (for removing stones from tyres)* ● *Washer fluid additive* ● *Small quantity of distilled water – for models fitted with batteries which require topping-up* ● *Pin (for adjusting washer nozzles)*

Note: *For details of how to identify the various engine types, refer to **'Servicing'** on page 103.*

REGULAR CHECKS

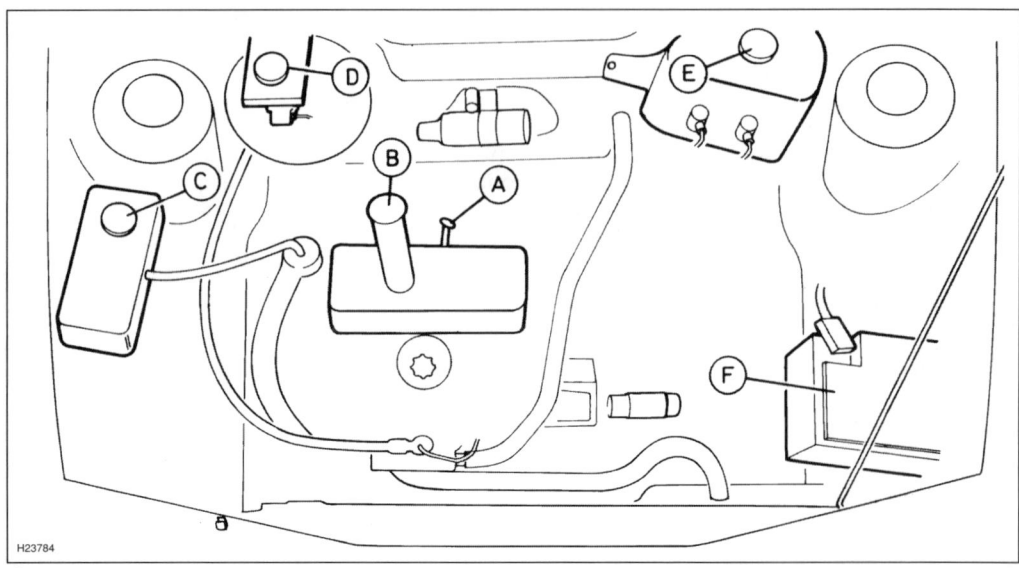

▲ Typical engine compartment layout – 1.3 litre & Maestro 1.6 litre 'R-series' engine models

A Engine oil level dipstick
B Engine oil filler cap
C Cooling system expansion tank

D Braking system fluid reservoir
E Windscreen (and rear window) washer fluid reservoir
F Battery

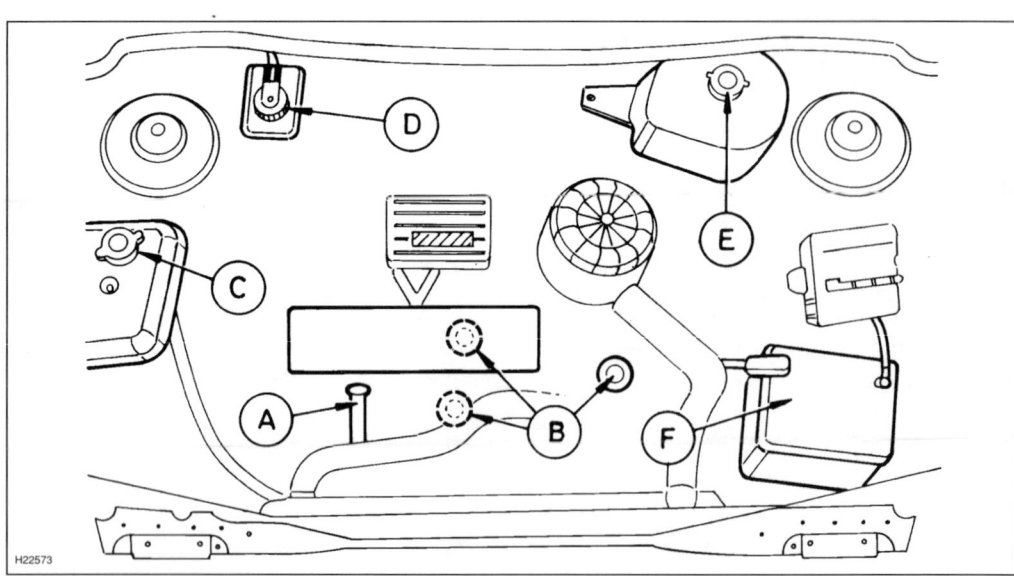

▲ Typical engine compartment layout – 1.6 litre 'S-series' & 2.0 litre models

A Engine oil level dipstick
B Engine oil filler cap (other locations also indicated)
C Cooling system expansion tank

D Braking system fluid reservoir
E Windscreen (and rear window) washer fluid reservoir
F Battery

AUSTIN/MG MAESTRO & MONTEGO

Checking oil level

To check the oil level, the car should be standing on level ground with the engine switched off. If the engine has just been run, wait for a few minutes to allow the level to settle before taking a reading.

Open the bonnet and look for the oil level dipstick, which has a plastic T-shaped top with the word 'OIL' marked on it. On all 1.3 litre models and early 1.6 litre Maestro models with the 'R-series' engine, the dipstick is at the rear of the engine, between the two middle spark plug leads. On all other 1.6 litre models, the dipstick is at the front right-hand (seen from the driver's seat) end of the engine. On all 2.0 litre models, the dipstick is at the front of the engine, between the two middle spark plug leads.

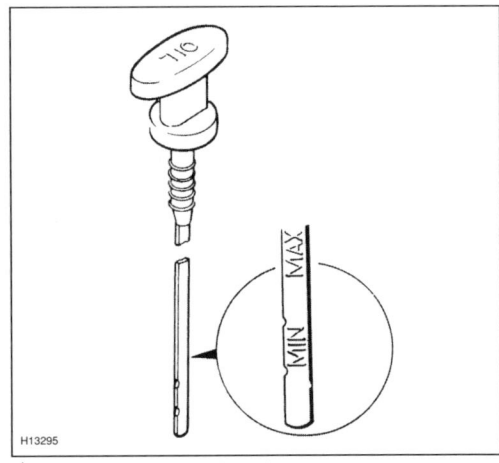
▲ Engine oil level dipstick markings

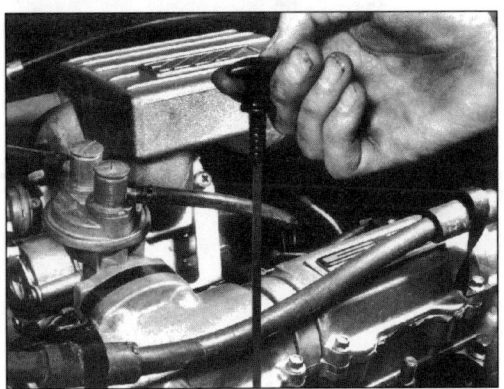
▲ Using dipstick to check level of engine oil – later 1.6 litre model shown

Pull out the dipstick and wipe it clean (with a clean, non-fluffy cloth). Slowly push the dipstick back into its tube until it seats fully, then pull it out again. Note the oil level, which should be between the 'MAX' and 'MIN' marks, and should never be allowed to drop below the 'MIN' mark.

If the oil needs topping-up, remove the oil filler cap. On all 1.3 litre models and early 1.6 litre Maestro models with the 'R-series' engine, the filler cap is located on the top of the engine; remove it by twisting it anti-clockwise to release and pulling it up. On 1.6 litre models from 1984 to 1988, the filler cap is located in a separate tube at the left-

hand (seen from the driver's seat) end of the engine, while on all 2.0 litre models, the filler cap is located in a separate tube at the front of the engine; remove it by pulling it out of the tube. On later 1.6 litre models, unscrew the filler cap from the left-hand (seen from the driver's seat) end of the engine's top cover.

Top-up to the dipstick 'MAX' mark; the amount of oil required to raise the level from the dipstick 'MIN' to 'MAX' marks is approximately 0.5 litre (0.88 pint). Always try to use the same type and make of oil, and take care not to overfill. Mop up any oil which might have been spilt, and ensure that the oil filler cap is correctly refitted.

If the oil needs to be topped-up regularly, check for leaks; if necessary, seek advice.

▲ Topping-up the engine oil

Checking coolant level

The coolant level must be checked by noting the level in the cooling system's expansion tank (which is located on the car's right-hand inner wing panel) when the engine is **cold**. Checking the level (and/or topping-up) means that the cooling system filler cap must first be removed from the expansion tank. The cap should be removed **only** when the engine is **cold**, but if it does prove necessary to remove the cap with the engine hot for any reason, take care to avoid scalding.

Whether the engine is hot or cold, protect yourself against possible injury by placing a thick rag over the filler cap, wearing gloves and keeping your face away from the tank opening; loosen the cap slowly and in stages, to gradually release the pressure in the system.

Remove the filler cap by slowly turning it anti-clockwise until it reaches its stop. Wait until any pressure in the system has escaped (when any hissing has stopped), then press the cap down and turn it further anti-clockwise; if any more hissing is heard, stop and wait for the sound to cease before removing the cap. Release the downward pressure slowly, and remove the cap.

Look inside the expansion tank; the coolant level should just cover the indicator post.

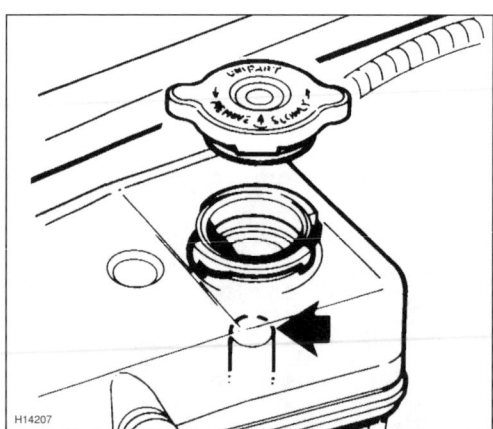

▲ *Expansion tank indicator post (arrowed) should just be covered if coolant level is correct*

Top-up the coolant (to just cover the indicator post) with a mixture of water and antifreeze to the same strength as the mixture already in the system (in this case, one-third antifreeze to two-thirds clean water). If there's any doubt about the strength of the coolant mixture, it's better to add too much antifreeze than not enough. It's important to note that because of the different types of metals used in the engine, it's vital to use antifreeze with suitable anti-corrosion additives all year round. Although plain water can be used for topping-up, it's unwise to make it a habit, as the strength of the antifreeze in the main system will gradually be diluted. **Never** use plain water to fill the whole system. Refit the filler cap, ensuring that it is securely fastened.

▲ *Topping-up the coolant*

Normally, topping-up will rarely be required; if regular topping-up is needed, it will probably be due to a leak somewhere in the system. Leaks are most likely to occur from the radiator or from the various hoses. If no leaks can be found, it's possible that there's an internal fault in the engine, such as a crack in the cylinder head or a blown cylinder head gasket, but in this case it's best to seek specialist advice.

Checking brake fluid level

Note: *Refer to* **'Safety first'** *on page 96 for the special precautions which should be taken when handling brake fluid.*

Although a low brake fluid level warning lamp is fitted, the fluid level should be checked visually whenever the oil and coolant levels are checked. The braking system fluid reservoir is located in the rear right-hand (seen from the driver's seat) corner of the engine compartment; the level of fluid can be seen through the reservoir's translucent body.

When the brake pads and shoes are all new, the level should be up to the 'MAX' mark on the side of the reservoir, but it is quite normal for the level to fall slightly as the brake friction material wears. The fluid level must **never** be allowed to drop below the 'MIN' mark; if it is ever found to be below the 'MIN' mark, it must be topped-up to this level.

When unscrewing the reservoir filler cap, first wipe it clean to prevent dirt from falling in when it is removed, then hold the terminal block at the centre of the cap still, to prevent damage to its connections, while unscrewing the cap's outer rim.

If topping-up is required, always use the correct type of fluid, which should always be stored in a full, airtight container. Don't top-up using fluid which has been stored in a partly-full container, as it will have absorbed moisture from the air, which can dangerously reduce its performance.

▲ *Topping-up the brake fluid level*

Topping-up should hardly ever be required, unless there's a leak somewhere in the hydraulic system. If a leak is suspected, the car should not be driven until the braking system has been thoroughly checked. **Never** take any risks where brakes are concerned.

Checking tyres

It's extremely important to carry out regular checks on the tyres, to make sure that the pressures are correct, and that the tyres are not damaged. The tyres are the only part of the car in contact with the road, so their condition will affect the steering and general handling of the car, and therefore its safety.

To check tyre pressures accurately, the tyres must be cold, which means that the car must not have been driven recently. Note that it can make quite a difference to the tyre pressures if the car has been standing out in the sun. This can be very noticeable when one side of the car is in the sun, and the other side is in shadow.

▲ *Checking a tyre pressure using a pencil-type gauge*

Note that the recommended tyre pressures vary depending on the size of tyres fitted – refer to *'Service specifications'* on page 80. The size of the tyre (eg 155 SR 13) is clearly marked on the tyre sidewall, although the style of the tyre size marking may vary depending on the tyre manufacturer – consult a Rover dealer or a tyre specialist for further details of tyre size markings, and for the latest pressure recommendations.

If the tyres are being checked after the car has been driven, for example when filling up with petrol during a journey, the pressures are bound to be higher than specified. It's best to simply check that the pressures in the two front tyres are *equal*, and similarly for the two back tyres (remember that the specified pressure for the rear tyres may be different to the front). **Never** let air out of the tyres of a car which has recently been driven, to bring the pressures down to that specified (unless the tyre pressures had been increased for a fully-loaded car, and the load has just been reduced).

When checking the tyre pressures, don't forget to check the spare!

Obviously, the spare should be kept at the highest pressure specified for the car; its pressure can always be reduced, if necessary, if it is ever needed, but pumping it up might not be so easy by the roadside!

Whenever the tyre pressures are checked, always refit the valve cap; this not only keeps dirt out of the valve, it also forms an additional seal against accidental deflation.

With the tyres properly inflated, run your fingers around the edge of the tyre to check for any cuts and bulges in the tyre walls. Ideally, the car should be jacked up to visually check the tyres all round but, alternatively, the car can be rolled backwards and forwards to enable you to check all round the tread. Check the tread for cuts, and remove any debris, such as small stones and broken glass, using a screwdriver or similar tool. Any large items such as nails which have penetrated the surface of the tyre should be left in the tread (to identify the location of the damage) until the tyre has been repaired. Refer to *'Breakdowns'* on page 49 for details of how to fit the spare wheel.

Also check the tyre treads for wear. Uneven wear across one particular tyre may be due to a fault with the suspension or steering. By law, the tread depth must be at least 1.6 mm (0.06 in), throughout a continuous band comprising the central three-quarters of the width of the tyre tread, around the full circumference of the tyre.

If you find that any of the tyres is excessively worn or damaged, obtain a new tyre as soon as possible. It's a good idea to try and stick to one type of tyre if possible, rather than mixing several different makes on the car. Generally, you'll find that it's much cheaper to buy tyres from a tyre specialist, rather than an ordinary garage. Make sure that when you buy a new tyre you have the wheel balanced, otherwise you might find that the new tyre causes vibration when driving.

Shoulder wear	Centre wear	Uneven wear	Toe wear
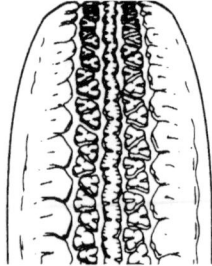	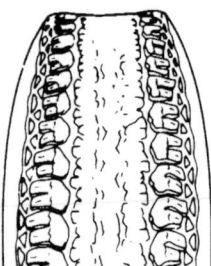	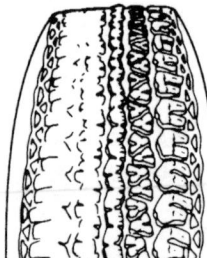	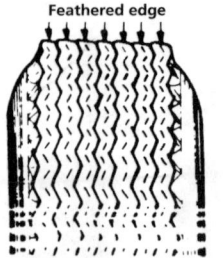

Shoulder wear

Probable cause:
Underinflation (wear on both sides)
Action: Check and adjust pressure

Probable cause:
Incorrect wheel camber (wear on one side)
Action: Repair or renew suspension parts

Probable cause:
Hard cornering
Action: Reduce speed

Centre wear

Probable cause:
Overinflation
Action: Measure and adjust pressure

Uneven wear

Probable cause:
Incorrect camber or castor
Action: Repair or renew suspension parts

Probable cause:
Malfunctioning suspension
Action: Repair or renew suspension parts

Probable cause
Unbalanced wheel
Action: Balance tyres

Probable cause:
Out-of-round brake disc/drum
Action: Machine or renew disc/drum

Toe wear

Feathered edge

Probable cause:
Incorrect toe setting
Action: Adjust front wheel alignment

▲ *Tyre wear patterns and causes*

Checking washer fluid level

The fluid reservoir for the windscreen (and, where fitted, the rear tailgate and/or headlamps) washer system is located in the rear left-hand (seen from the driver's seat) corner of the engine compartment.

▲ Topping-up the washer reservoir

If topping-up is necessary, remove the reservoir cap and top-up as necessary with clean water. A suitable washer fluid additive will keep the glass free from smears, and will prevent the fluid freezing in Winter; follow the manufacturer's instructions as to mixture strength. **Do not** use engine antifreeze in a washer system, as it will damage the car's paintwork, the window glass rubber seals and the wiper blades.

Checking battery electrolyte level

All models were originally fitted with a maintenance-free battery during production. The maintenance-free battery is 'sealed for life', and does not require topping-up with distilled water. A maintenance-free battery is usually clearly marked, and will have no removable covers to enable topping-up.

Where the original battery has been replaced by a standard or low-maintenance type battery, the electrolyte level should be checked as follows.

The battery is located in the front left-hand (seen from the driver's seat) corner of the engine compartment. Remove the caps or the cover from the top of the battery, and check the level of the electrolyte fluid (with some

types of battery you can see the fluid level through the battery case). The level should be just above the tops of the metal plates inside the battery, or in some cases up to a mark on the battery case.

If necessary, add distilled water (not ordinary tap water) to each cell, to bring the level just above the tops of the battery plates, or up to the marks, as applicable.

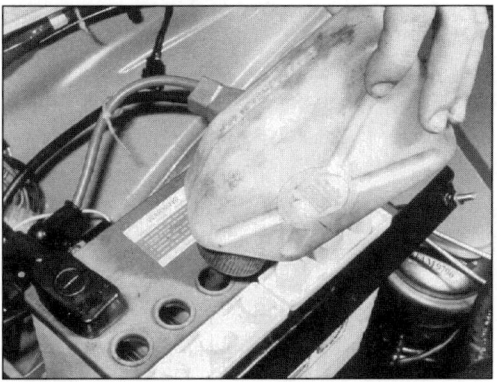

▲ Topping-up the battery electrolyte level

With some types of battery, distilled water is added to a trough in the top of the battery until all the filling slots are full, and the bottom of the trough is just covered. On completion, refit the caps or cover, and carefully wipe up any drops of water that were spilt.

Topping-up should hardly ever be required, and the need for frequent topping-up indicates that the battery is being overcharged due to a fault in the charging circuit – seek advice from someone suitably qualified if this is the case.

Checking wipers and washers

Check all the wipers and washers to make sure that they're working properly. Don't allow the wipers to work on dry glass for too long, as it will strain the motor and wear out the wiper blades. Blocked washer nozzles can be cleared using a piece of thin wire as a probe. If necessary, the windscreen (and Montego Estate rear window) washer nozzles can be adjusted using a pin inserted into the end of the nozzle; Montego headlamp washer nozzles, and Maestro rear window washer nozzles, are fixed. If you're adjusting the windscreen washer nozzles, remember to aim them fairly high on the windscreen, as the airflow will usually deflect the spray down when the car is moving; Montego Estate rear window washer nozzles should be set to direct the spray to the tip of the wiper blade, on each side.

Over a period of time, the wiping action of the wiper blades will deteriorate, causing smearing of the glass. The rubber may also crack, particularly at the edges of the blades. When this happens, the blades must be renewed; note that Rover recommend that the blades should be renewed annually, as a matter of course, to ensure that they never get to this state.

Blade renewal is a straightforward job, and is accomplished as follows:

Pull the wiper arm away from the glass, against the spring pressure, until the arm clicks into position.

On early Maestro models, lift the small retaining lever on the end of the arm, and separate the wiper blade from the arm.

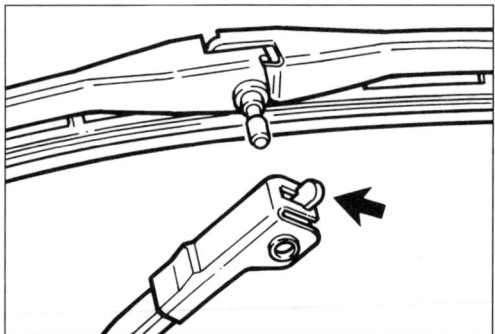

▲ Lift lever (arrowed) to release wiper blade from arm - early Maestro models

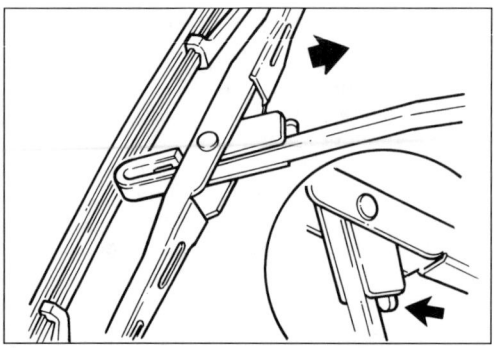

▲ Press lever (arrowed), then slide blade as shown to release – all other models

On all other models, press the small retaining lever on the blade, and slide the blade down the arm until it can be unhooked.

It is usual to renew the complete blade as an assembly, but note that replacement rubber inserts may be available from some accessory shops; these may prove cheaper than complete blades, but fitting is often fiddly. Follow the manufacturer's instructions to fit them.

Fit the new blade by reversing the removal procedure; ensure that the blade is secured correctly by its fastening.

Checking lights and horn

Switch on all the lights in turn, and check that they're working. Don't forget to check the direction indicators and the brake lamps (the ignition must be switched on to check these), either with the help of an assistant, or by looking for the reflection in a suitable window or door etc.

Check that the direction indicators work with the brake lamps on, and that the brake lamps work with the tail lamps on. Some faults may cause the various rear lamps to interact, which can be dangerous, as it may confuse or distract following drivers.

Check the operation of the horn. A short 'blast' should be enough to prove that it at least works, but from time to time check its operation over several seconds.

If any of the bulbs need renewing, refer to 'Bulb, fuse and relay renewal' on page 127. Any other problems are likely to be caused by a faulty switch, or loose or corroded connections (especially earth connections), and it's best to seek advice to solve these.

Checking for fluid leaks

Check the ground where the car is normally parked for any stains or spots of fluid which may have been caused by fluid leaking from the car.

Open the bonnet, and make a quick check of all the hoses and pipes, and the surfaces of all the components in the engine compartment. If there's any sign of leaking fluid, try to find the source of the leak, and seek advice if necessary. It can be difficult to identify leaking fluids, but if there's obviously a major leak, or if you suspect even a slight brake fluid or petrol leak, don't drive the car until the problem has been investigated by someone suitably qualified.

Regular servicing will ensure that your car is reliable and safe to drive, and could save you a lot of money in the long run. Many of the unexpected expenses which can crop up if things go wrong with your car can be avoided by carrying out regular servicing.

A significant proportion of the average garage servicing bill (well over 50% in many cases) is made up of labour costs, so obviously a lot of money can be saved by carrying out the work yourself. You'll find that most of the servicing jobs are very straightforward, and you don't need to be a mechanical genius to 'have a go' yourself. Don't be put off when you look under your bonnet; there are surprisingly few items which require frequent attention, and most of those which do are easily accessible without the need for anything more than basic tools. You'll discover that DIY servicing can be very rewarding, and will help you to understand what makes your car 'tick'.

The following chart lists all the servicing tasks recommended by the car manufacturer, and details of how to carry out the necessary work can be found in the subsequent pages. A few of the jobs require more extensive knowledge, or the use of special tools, and detailed explanation is beyond the scope of this Handbook. These more complicated tasks are identified on the chart, and details can be found in our Owners Workshop Manual for your particular model. Even if you decide to have the work done by a garage, the chart will enable you to check that the necessary work has been carried out.

Whenever you're carrying out servicing, safety must always be the first consideration, and you should read through the *'Safety first!'* notes on page 96 before proceeding any further.

SERVICING

ENGINE COMPARTMENT

Check all components for fluid leaks, corrosion or deterioration

ENGINE

Check engine oil level

Check for oil leaks

Renew engine oil & filter

Renew engine oil filler cap – 1.3 litre models

Renew engine breather filter – Maestro 1.6 litre models with 'R-series' engine

Clean engine breather filter – all other 1.6 litre models & 2.0 litre models

Check valve clearances – 1.3 litre models

Check valve clearances – 1.6 & 2.0 litre models

COOLING SYSTEM

Check coolant level

Check hoses, clips & joint surfaces

Check, and adjust or renew, alternator/water pump drivebelt

Renew coolant

FUEL AND EXHAUST SYSTEMS

Check condition of fuel pipes & hoses

Check condition of fuel tank

Check condition of exhaust system

Top-up carburettor piston damper

Check operation of accelerator & linkage (and manual choke, where fitted)

Renew air cleaner filter element – all Maestro models, Montego 2.0 litre models

Renew air cleaner filter element – Montego 1.3 & 1.6 litre models

Check idle speed & mixture settings

Have the carburettor balance checked – MG Maestro 1600 models

Renew fuel filter – carburettor models, where fitted

Note: *In addition to the tasks listed in this table, Rover specify the following:*
On Turbo models, the engine oil and filter must be renewed, and the carburettor piston damper must be topped-up, every 6000 miles (10 000 km) or six months, whichever occurs first; work as described for all other 2.0 litre models in the 12 000 mile (20 000 km) service section on page 108.
On all fuel-injected 2.0 litre models, the fuel filter must be renewed every 48 000 miles (80 000 km) or 4 years, whichever comes first; this task should be referred to a Rover dealer.

| Regular | 12 000 miles | 24 000 miles |
| Weekly | (20 000 km) | (40 000 km) |
	1 year	2 years
■	■	■
■	■	■
	■	■
	■	■
		■
		■
		■
	□	□
		□
■	■	■
	■	■
	■	■
		■
	■	■
	■	■
	■	■
	■	■
	■	■
		■
	□	□
		□
		□

Items marked □ are considered to be beyond the scope of this Handbook, and details can be found in the Owners Workshop Manual for your car (OWM 922 for Maestro 1.3 & 1.6 models, OWM 1066 for Montego 1.3 & 1.6 models or OWM 1067 for Montego 2.0 models). Note that these manuals are continually being updated – check with your local stockist to ensure that the latest-available edition is obtained.

AUSTIN/MG MAESTRO & MONTEGO

SERVICE SCHEDULE 1

IGNITION SYSTEM

Clean distributor cap, ignition coil & HT leads

Renew spark plugs – Montego 1.3 & 1.6 litre models

Renew spark plugs – all Maestro models, Montego 2.0 litre models

Check ignition timing – 1.3 litre models & Maestro 1.6 litre models with 'R-series' engine

Lubricate distributor – 1.3 litre models & Maestro 1.6 litre models with 'R-series' engine

MANUAL GEARBOX

Check operation of clutch & pedal

Check for oil leaks

Check condition of driveshaft joints & rubber gaiters

Check, and adjust if necessary, clutch cable

Check oil level

Renew gearbox oil – 2.0 litre models

Lubricate gearchange linkage – 1.3 & 1.6 litre models

AUTOMATIC TRANSMISSION

Check for fluid leaks

Check fluid level

Check condition of driveshaft joints & rubber gaiters

Check operation of parking pawl

Check final drive oil level – 1.6 litre models

Renew fluid

BRAKING SYSTEM

Check brake fluid level

Check condition of pipes, flexible hoses & unions

Check operation of warning lamps fitted

Check condition & security of servo vacuum hose

Check brake operation

Check front brake pads & discs

Check rear brake shoes & drums

Renew hydraulic fluid

Note: *In addition to the tasks listed in this table, Rover specify the following:*

On 1.6 litre models with the 'S-series' engine, and all 2.0 litre models, the timing belt must be renewed every 48 000 miles (80 000 km) or 4 years, whichever comes first; this task should be referred to a Rover dealer.

On all models, the braking system must be overhauled, with all flexible hoses, the servo air filter and vacuum hose non-return valve, and all master cylinder/brake caliper/wheel cylinder seals being renewed, every 36 000 miles (60 000 km) or 3 years, whichever comes first; this task should be referred to a Rover dealer.

Regular	12 000 miles (20 000 km)	24 000 miles (40 000 km)
Weekly	1 year	2 years
	■	■
	■	■
		■
		□
		□
	■	■
	■	■
	■	■
	□	□
	□	□
	□	□
	□	□
	■	■
	■	■
	■	■
		■
	□	□
		□
■	■	■
	■	■
	■	■
	■	■
	■	■
	□	□
	□	□
		□

Items marked □ are considered to be beyond the scope of this Handbook, and details can be found in the Owners Workshop Manual for your car (OWM 922 for Maestro 1.3 & 1.6 models, OWM 1066 for Montego 1.3 & 1.6 models or OWM 1067 for Montego 2.0 models). Note that these manuals are continually being updated – check with your local stockist to ensure that the latest-available edition is obtained.

AUSTIN/MG MAESTRO & MONTEGO

SERVICE SCHEDULE 2

SUSPENSION AND STEERING

Check tyres for condition & pressure

Check tightness of roadwheel nuts, and check wheels for damage

Check power-assisted steering fluid level

Check, and adjust or renew, power-assisted steering pump drivebelt

Check front & rear suspension struts for fluid leaks

Check condition and security of steering gear & suspension components, balljoints & rubber gaiters

If the front tyres show signs of uneven wear, have the wheel alignment checked

INTERIOR AND BODYWORK

Check condition and security of seats & seat belts

Lubricate all hinges, locks and the bonnet release mechanism

Check condition of underseal

Inspect paintwork & bodywork

ELECTRICAL SYSTEM

Check washer fluid level

Check battery electrolyte level (where applicable)

Check wipers & washers

Check operation of all lights & electrical equipment

Check condition of accessible wiring connectors, harness & clips

Clean battery terminals

Have the headlamp aim checked

CAR UNDERSIDE

Check all components for fluid leaks, corrosion or deterioration

ROAD TEST

Check instruments & electrical equipment

Check steering, suspension & general handling

Check engine, clutch (where applicable), gearbox & driveshafts

Check braking system

	Regular	12 000 miles (20 000 km)	24 000 miles (40 000 km)
	Weekly	1 year	2 years
	■	■	■
		■	■
		■	■
		■	■
		■	■
		□	□
		□	□
		■	■
		■	■
		■	■
		■	■
	■	■	■
	■	■	■
	■	■	■
	■	■	■
		■	■
		■	■
		■	■
		■	
		■	■
		■	■
		■	■
		■	■

SERVICE SCHEDULE 3

Items marked □ are considered to be beyond the scope of this Handbook, and details can be found in the Owners Workshop Manual for your car (OWM 922 for Maestro 1.3 & 1.6 models, OWM 1066 for Montego 1.3 & 1.6 models or OWM 1067 for Montego 2.0 models). Note that these manuals are continually being updated – check with your local stockist to ensure that the latest-available edition is obtained.

SAFETY FIRST!

Professional motor mechanics are trained in safe working procedures. No matter how enthusiastic you may be about getting on with the job you have planned, take time to read this Section. Don't risk an injury by failing to follow the simple safety rules explained here. The following is a guide to encourage you to be 'safety conscious' as you work on your car.

Essential points

ALWAYS use a safe system of supporting your car when working underneath it. The car's own jack (or a single hydraulic jack) is never sufficient. Once you've raised the car, you should use a safe means of holding it up. Axle stands, ramps or substantial wooden blocks are good; concrete blocks and bricks may crack or disintegrate, and should not be used. Make sure that you locate the supports where you know the car won't collapse, and where they can't slip.

ALWAYS loosen wheel nuts and other 'high torque' nuts or bolts (as applicable for the task to be carried out) before jacking up your car, or it may slip and fall.

ALWAYS make sure that your car is in **Neutral** or in **Park** (in the case of automatic transmission), and that the handbrake is securely on before trying to start the engine.

UNLESS the engine is cold, **ALWAYS** cover the cooling system's pressure (expansion tank) cap with several thicknesses of cloth before trying to remove it. Release the pressure slowly or the coolant may escape suddenly and scald you.

ALWAYS wait until the engine has cooled down before draining the engine (and/or transmission) oil, or the coolant. Oil or coolant that is very hot may scald you. If you can touch the engine sump/transmission/radiator (as applicable) without discomfort, the engine has probably cooled sufficiently to avoid scalding.

ALWAYS allow the engine to cool before working on it. Many parts of the engine become very hot in normal operation, and you may burn yourself badly.

ALWAYS keep brake fluid, antifreeze and other similar liquids away from your car's paintwork, as it may damage the finish (or may remove the paint altogether!).

ALWAYS keep toxic fluids, such as fuel, brake and transmission fluids and antifreeze off your skin, and never syphon them by mouth.

ALWAYS wear a mask when doing any dusty work, or where spraying is involved, especially when doing body repair and brake jobs.

ALWAYS clean up oil and grease spills – someone may slip and be injured.

ALWAYS use the right tool for the job. Spanners and screwdrivers which don't fit properly are likely to slip and cause injury.

ALWAYS get help to lift and handle heavy parts – it's never worth risking an injury.

ALWAYS take time over the jobs you take on. Plan out what you must do, and make sure that you have the right tools and spare parts. Follow the recommended steps, and check over the job once you're finished. Leave 'short-cuts' to the experts.

ALWAYS wear eye protection when using power tools such as drills, sanders and grinders, when using a hammer, and when working beneath your car.

ALWAYS use a barrier cream on your hands when doing dirty jobs, especially when in contact with fuel, oils, greases, and brake and transmission fluids. It will help protect your skin against infection, and will make the dirt easier to remove later. Make sure that your hands aren't slippery. The use of a suitable specialist hand cleaner will make dirt and grease easier to remove without causing infection or damage to skin. Long term or regular contact with used oils and fuel can be a health hazard.

ALWAYS keep loose clothing and long hair out of the way of any moving parts.

ALWAYS take off rings, watches, bracelets

and neck chains before starting work on your car, especially the electrical system.

ALWAYS make sure that any jacking equipment or lifting tackle you use has a safe working load which will cope with what you intend to do, and use the equipment exactly as recommended by the manufacturers.

ALWAYS keep your work area tidy; someone may trip or slip on articles left lying around and be injured.

ALWAYS get someone to check up on you from time to time if you're working on the car alone, to make sure that you're alright.

ALWAYS do the work in a logical order, check that you've put things back together properly, and that everything is tightened as it should be.

ALWAYS keep children and animals away from an unattended car, and from the area where you're working.

ALWAYS park cars with catalytic converters away from materials which may burn, such as dry grass, oily rags, etc, if the engine has recently been run. Catalytic converters reach extremely high temperatures, and any such materials close by may catch fire.

REMEMBER that your safety, and that of your car and other people, rests with you. If you are in any doubt about anything, get specialist advice and help right away.

IF in spite of these precautions you injure yourself – get medical help immediately.

Asbestos

Some parts of your car, such as brake pads and linings, clutch linings and gaskets contain asbestos, and you should take appropriate precautions to avoid inhaling dust (which is hazardous to health) when working with them If in doubt, assume that they **do** contain asbestos.

Fire

Remember at all times that petrol is highly flammable. Never smoke, or have any kind of naked flame around, when working on the car. But the risk doesn't end there – a spark caused by an electrical short-circuit, by two metal surfaces contacting each other, by careless use of tools, or even by static electricity built up in your body under certain conditions, can ignite petrol vapour, which in a confined space is

highly explosive.

If a fuel leak is suspected, try to find the cause; seek advice as soon as possible, and never risk fuel leaking onto a hot engine or exhaust. Catalytic converters (where fitted) run at extremely high temperatures, and therefore can be an additional fire hazard – observe the precautions outlined previously in this Section.

It's recommended that a fire extinguisher of a type suitable for fuel and electrical fires is kept handy in the garage or workplace at all times.

Never try to extinguish a fuel or electrical fire with water.

Fumes

Certain fumes are highly toxic, and can quickly cause unconsciousness and even death if inhaled to any extent, especially if inhalation takes place through a lighted cigarette or pipe. Petrol vapour comes into this category, as do the vapours from certain solvents such as trichloroethylene. Any draining or pouring of such fluids should be done in a well-ventilated area.

When using cleaning fluids and solvents, read the instructions carefully. Never use materials from unmarked containers – they may give off poisonous vapours.

Never run a car engine in an enclosed space such as a garage. Exhaust fumes contain carbon monoxide which is extremely poisonous; if you need to run the engine, always do so in the open air, or at least have the rear of the car outside the workplace. Although cars fitted with catalytic converters produce far less toxic exhaust gases, the above precautions should still be observed.

If you're fortunate enough to have the use of an inspection pit, never run the engine while the car is standing over it; the fumes, being heavier than air, will concentrate in the pit with possibly lethal results.

The battery

Never short across the two poles of the battery! (A battery is shorted by connecting the two terminals directly to each other, and this can happen accidentally when working under the bonnet with metal tools.) The heavy discharge caused will create 'gassing' of hydrogen from the battery, and this is highly explosive. Shorting across a battery may also cause sparks, and the combination can cause the battery to explode, with potentially lethal results. A conventional battery will normally be giving off a certain amount of hydrogen all the time, so for the same reasons given above, never cause a spark or allow a naked flame close to it.

Batteries which are sealed for life require special precautions which are normally outlined on a label attached to the battery. Such precautions usually relate to battery charging and jump starting from another vehicle (for details of jump starting refer to 'Breakdowns' on page 56).

If possible, loosen the filler plugs or covers when charging the battery from an external source. Don't charge at an excessive rate, or the battery may burst. Special care should be taken with the use of high charge-rate boost chargers to prevent the battery from overheating.

Take care when topping-up and when carrying the battery. The battery contains dilute sulphuric acid which is very corrosive. It will burn and may cause long-term damage if in contact with skin, eyes or clothes. Similarly, corrosive deposits around the battery terminals may be harmful.

Always wear eye protection when cleaning the battery, to prevent the corrosive deposits from entering your eyes.

Mains electricity and electrical equipment

When using any electrical equipment which works from the mains, such as an electric drill or inspection light, always ensure that the appliance is correctly connected to its plug and that, where necessary, it's properly earthed. Always use an RCD (Residual Current Device, a safety device incorporating a circuit breaker) when using mains electrical equipment. Don't use such appliances in damp conditions, and take care not to create a spark or apply excessive heat close to fuel or fuel vapour. Also make sure that the appliances meet the relevant national safety standards.

Ignition HT voltage

A severe electric shock can result from touching certain parts of the ignition system, such as the spark plug HT leads, when the engine is running or being cranked, particularly if components are damp or the insulation is faulty. Where an electronic ignition system is fitted, the HT voltage is much higher than that used in a contact breaker system, and could prove fatal, especially to wearers of cardiac pacemakers.

Disposing of used engine oil

Used engine oil is a hazard to health and the environment, and should be disposed of safely and cleanly. Most local authorities provide a disposal site which will have a special tank for waste oil. If in doubt, contact your local council for advice on where you can dispose of oil safely – there may be a local garage who will allow you to use their specialised oil disposal tank free of charge. Remember that it's not just the oil which causes a problem, but empty containers, old oil filters, and oily rags, so these should be taken to your local disposal site too.

Do not under any circumstances pour engine oil down a drain, or bury it in the ground.

Jacking and vehicle support

The jack provided with the car is designed for emergency wheel changing, and should not be used for servicing and overhaul work. Instead, a more substantial workshop jack (trolley jack or similar) should be used. Whichever type is used, it's essential that additional safety support is provided by means of axle stands designed for this purpose. Never use makeshift means such as narrow wooden blocks or piles of house bricks, as these can easily topple or, in the case of bricks, disintegrate under the weight of the car. Further information on the correct positioning of the jack and axle stands is provided at the end of this Section.

If you don't need to remove the wheels, the use of drive-on ramps is recommended. Ensure that the ramps are correctly aligned with the wheels, and that the car is not driven too far along them, so that it promptly falls off the other end, or tips the ramps.

JACK AND AXLE STAND POSITIONS

When using a trolley jack or other type of workshop jack to raise one side of the car (at the front or rear), place the jack head under the relevant sill jacking point (**2** or **3**) with axle stands placed under the nearest appropriate jacking bracket/point or body member (**1**, **2**, **3** or **4**).

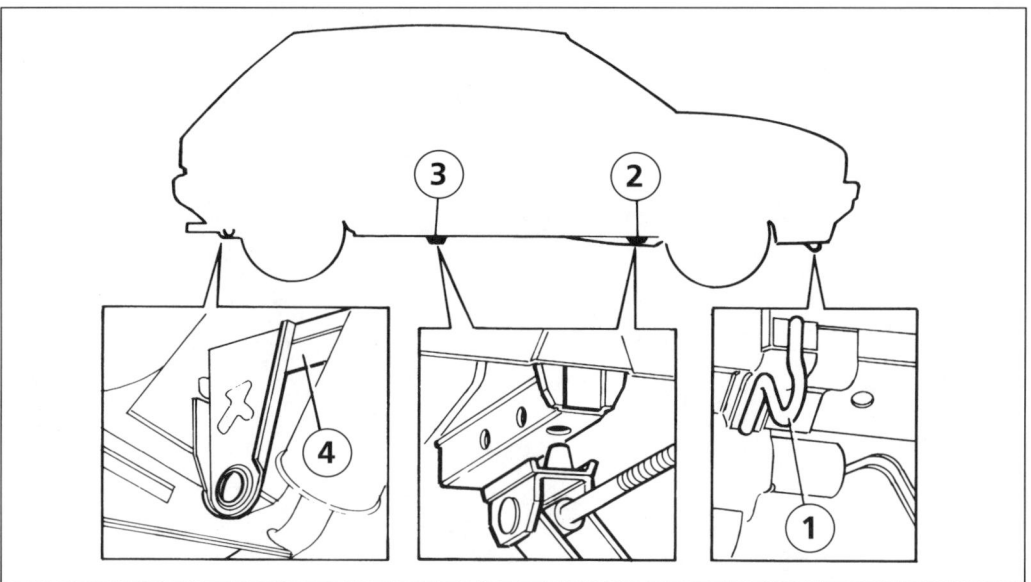

▲ *Jacking and support points on car underside – Maestro shown, Montego similar*
  1 *Front jacking bracket/towing eye*
  2 *Front sill (roadside) jacking points/axle stand locations*
  3 *Rear sill (roadside) jacking points/axle stand locations*
  4 *Body rear longitudinal member – one each side*

To raise the front of the car evenly, the jack head should be placed under the front jacking bracket/towing eye **1,** with axle stands placed under both sill jacking points **2**.

To raise the rear of a Maestro evenly, the jack head should be placed under the centre of a jacking beam (a suitable length of heavy square-section steel tubing) placed across the rear ends of the rear axle, under the suspension strut bottom mountings, with axle stands located under the sill jacking points **3**.

To raise the rear of a Montego evenly, the jack head should be placed under the rear jacking bracket/towing eye, with axle stands located under the sill jacking points **3**.

Never jack up under any (other) part of the suspension, or under any of the steering and transmission components.

If the engagement of the jack head or axle stands with the car's body is not precise and may result in the car being raised or supported insecurely, use a piece of wood to spread the load, shaped if necessary to clear any surface obstructions.

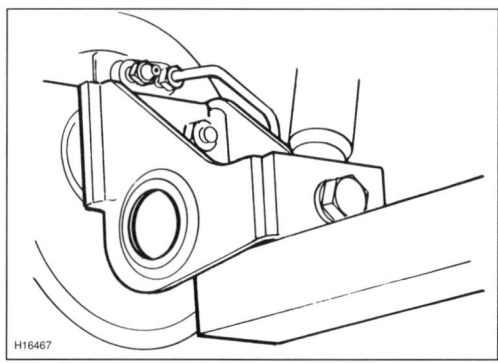

▲ *Location of jacking beam to raise rear of Maestro*

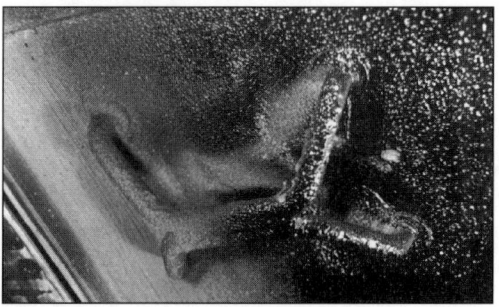

▲ *Rear jacking bracket/towing eye – Montego*

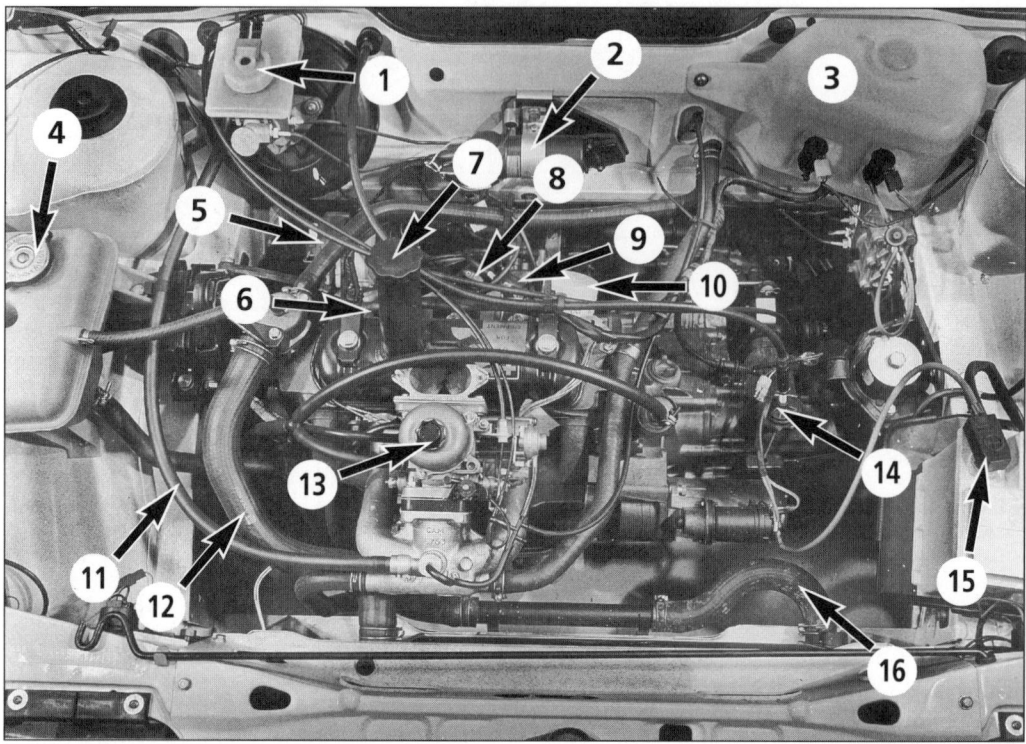

▲ *Underbonnet component locations – 1.3 litre model (air cleaner removed for clarity)*

1   Braking system master cylinder and fluid reservoir
2   Ignition coil
3   Windscreen (and rear window) washer fluid reservoir
4   Cooling system expansion tank with
     filler/pressure cap
5   Alternator
6   Number 1 cylinder spark plug HT lead
7   Engine oil filler cap
8   Engine oil level dipstick

9   Distributor cap and HT leads
10   Engine oil filter
11   Braking system servo vacuum hose
12   Radiator top hose
13   Carburettor piston damper
14   Clutch cable
15   Battery positive terminal (not earth)
16   Radiator bottom hose

▲ *Underbonnet component locations – 1.6 litre model (early 'S-series' engine shown)*

1 Braking system master cylinder and fluid reservoir
2 Braking system servo vacuum hose
3 Clutch cable
4 Carburettor piston damper
5 Ignition coil
6 Air cleaner
7 Windscreen (and rear tailgate) washer fluid reservoir
8 Engine oil filter
9 Cooling system expansion tank with
   filler/pressure cap

10 Vehicle Identification Number (VIN) plate
11 Number 1 cylinder spark plug HT lead
12 Alternator
13 Engine oil level dipstick
14 Radiator top hose
15 Distributor cap and HT leads
16 Engine oil filler cap
17 Battery positive terminal (not earth)
18 Radiator bottom hose

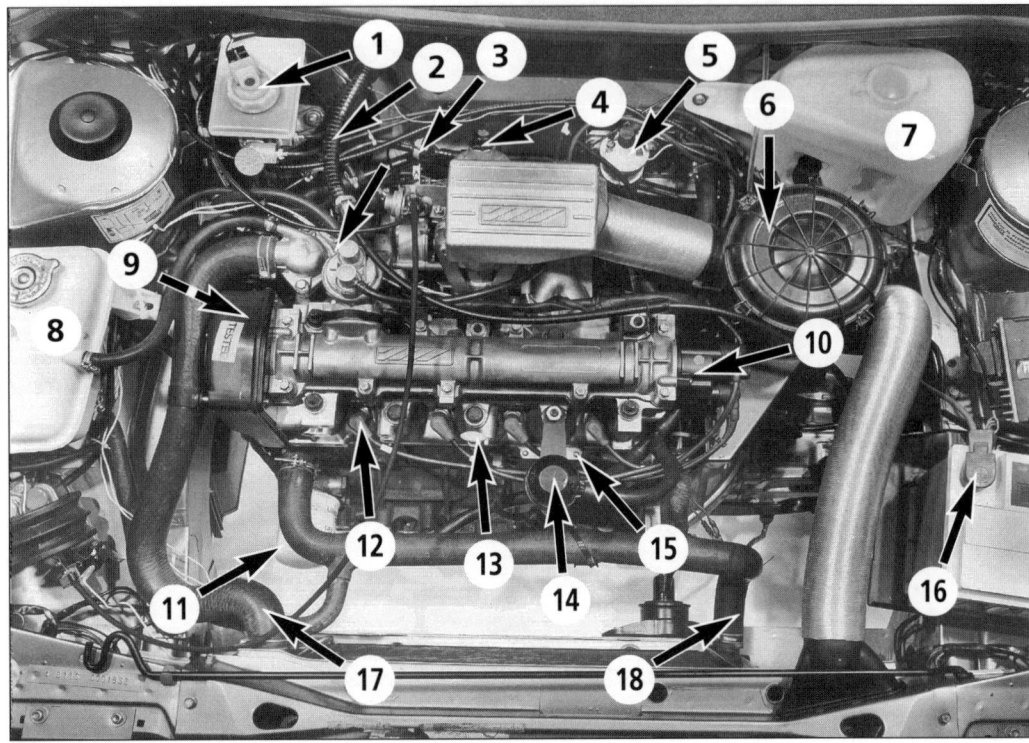

▲ *Underbonnet component locations – 2.0 litre model (early carburettor engine shown)*

1	Braking system master cylinder and fluid reservoir	**10**	Distributor cap and HT leads
2	Clutch cable	**11**	Engine oil filter – at rear of engine on later models
3	Braking system servo vacuum hose	**12**	Number 1 cylinder spark plug HT lead
4	Carburettor piston damper	**13**	Engine oil level dipstick
5	Ignition coil	**14**	Engine oil filler cap
6	Air cleaner	**15**	Engine number plate
7	Windscreen (and rear window) washer fluid reservoir	**16**	Battery positive terminal (not earth)
8	Cooling system expansion tank with filler/pressure cap	**17**	Radiator top hose
9	Alternator (location indicated) – at front of engine on later models	**18**	Radiator bottom hose

BUYING SPARE PARTS

When buying spare parts such as an oil or air filter, make sure that the correct replacement parts are obtained for your particular car. Many changes are made during the production run of any car, and when ordering spare parts, it will usually be necessary to know the year the car was built, the model type (eg 'MG EFi' or 'City') and engine size.

In some cases, however, further information will be required; a Rover dealer may need to know the Vehicle Identification Number (VIN) and the Engine Number before the correct replacement part can be ordered.

The VIN is stamped on a metal plate; on models up to 1986, this is attached to the front right-hand (seen from the driver's seat) corner of the engine compartment, while on all later models it is attached to the left-hand front door rear pillar, except for MG Montego Turbo models, where it is attached to the bonnet locking panel at the front of the engine compartment. The VIN is repeated, stamped into the centre of the drain channel at the rear of the bonnet on Maestro models, or into the right-hand front suspension strut top mounting on Montego models.

The engine number is either stamped into the cylinder block, or onto a metal plate attached to the cylinder block; in either case, it will be found on the top surface of the cylinder block, underneath the spark plugs.

Both numbers will also appear on the Vehicle Registration Document; check that they are correctly printed and always quote them in full, with their prefixes.

All of the components required for servicing will be available from an official Rover dealer, but most parts should also be available from good motor accessory shops and motor factors, possibly at lower prices.

SERVICE TASKS

The following pages provide easy-to-follow instructions to enable you to carry out the regular service tasks recommended by the manufacturer.

Identifying engines

All 1.3 litre models are fitted with the 'A+' engine, and all 2.0 litre models have the 'O-series' unit. Note, however, that the 'O-series' engine was significantly modified from late 1985 onwards (engine number prefix 20HB or subsequent), the principal changes being the repositioning of the alternator and oil filter. Note also that all 2.0 litre models are fitted with either a carburettor (with turbocharger, on MG Turbo models) or one of two fuel injection systems.

All 1.6 litre models use the 'S-series' engine, except for early (1983/4) Maestros, which are fitted with the 'R-series' engine. The 'R-series' engine can be identified by the distributor being on the rear of the engine's right-hand end (seen from the driver's seat). The distributor has a black plastic cap, connected by the five spark plug/HT leads to the four spark plugs and to the ignition coil. The finned aluminium alloy cover at the top of the R-series engine (below and to the rear of the air cleaner assembly) is in one piece. On the 'S-series' engine (seen from the driver's seat) a black plastic cover is fitted to its top right-hand end, the distributor is on the top left-hand end, and two separate aluminium alloy covers are bolted in between.

Every 250 miles (400 km) or weekly

Refer to *'Regular checks'* on page 81 and carry out all the checks described.

Every 12 000 miles (20 000 km) or 1 year – whichever comes first

ENGINE COMPARTMENT

Check all hoses and pipes, and the surfaces of all components for signs of fluid leakage, corrosion or deterioration

If there's any sign of leaking fluid, try to find the source of the leak, and seek advice if necessary.

It can be difficult to identify leaking fluids, but if there's obviously a major leak, or if you suspect even a slight brake fluid or petrol leak, don't drive the car until the problem has been investigated by someone suitably qualified.

Check all rubber or fabric hoses for signs of cracking, damage due to rubbing on other components, and general deterioration. It's sensible to have any suspect hoses renewed as a precaution against possible failure.

Check for obvious signs of corrosion on metal pipes, hose clips, and component joints, which may indicate a fluid leak. Any badly-corroded components should be cleaned and, if necessary, renewed. Pay particular attention to metal fluid pipes, and have them renewed if they're badly pitted or corroded.

ENGINE

Check for oil leaks

Refer to the *'ENGINE COMPARTMENT'* checks above.

Renew engine oil and filter

Note: *The following items will be required for this task:*
- *A suitable quantity of engine oil of the correct type – refer to* **'Service specifications'** *on page 77*
- *New oil filter (of the correct type for your particular car)*
- *New oil drain plug sealing washer (depending on the condition of the old one)*
- *Suitable container to catch the old oil as it drains (make sure that the capacity of the container is larger than the oil capacity of the engine – refer to* **'Service specifications'** *on page 78).*
- *Oil filter wrench – for removal of old oil filter*
- *Small quantity of clean rag*
- *Suitable spanner or socket to fit oil drain plug*

The oil should be drained when the engine is warm, immediately after a run. If you can touch the sump (the large pressed steel container or aluminium alloy casting at the bottom of the engine, visible behind and below the front body panel) without discomfort, the oil has probably cooled sufficiently to avoid scalding.

On 1.3 litre models, there is no need to jack up the front of the car to improve access; just be careful not to bark your knuckles on the ground when unscrewing the drain plug! On all other models, since the drain plug is at the rear of the sump, draining will be quicker and more effective (and drain plug/oil filter access will be improved on most cars) if the front of the car is first jacked up – if the car is to be raised, apply the handbrake firmly and chock the rear wheels, then jack up the front of the car and support it securely on axle stands (refer to *'Jacking and vehicle support'* in *'Safety first!'* on page 98 for details of where to position the jack and stands).

Position a suitable container under the drain plug to catch the oil; on 1.3 litre models, and early Maestro 1.6 litre models with the 'R-series' engine, the drain plug is in the right-hand end (seen from the driver's seat) of the sump, while on all other models it is at the rear of the sump, towards its right-hand end (seen from the driver's seat).

▲ *Location (arrowed) of engine oil drain plug – 1.3 litre models ...*

▲ *... early Maestro 1.6 litre models – 'R-series' engine ...*

▲ ... all other models – typical

▲ Location of engine oil filter – 1.3 litre models

Remove the oil filler cap from the top of the engine, then working under the car, unscrew the drain plug using a suitable spanner or a socket. Oil will be released before the drain plug is removed completely, so take precautions against scalding, as the oil will be hot. Position the container to allow for the fact that the initial jet of oil may fall at least eight inches away from the drain hole, and try not to drop the drain plug when removing it.

Allow the oil to drain until the flow has stopped. Check the condition of the oil drain plug sealing washer, and renew it if necessary. If a magnetic insert is fitted in the tip of the drain plug, wipe it clean.

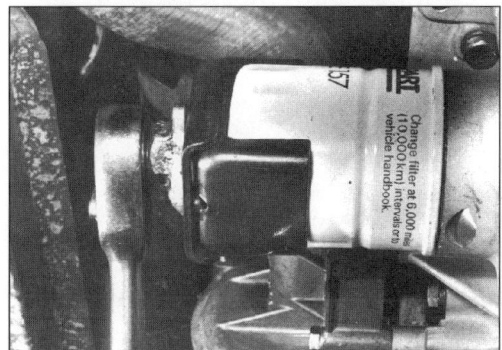

▲ Using a typical filter wrench to unscrew 'R-series' engine oil filter – early Maestro 1.6 litre model

When the oil has finished draining, clean the drain plug, washer, and the sump mating face, then refit and tighten the drain plug. When tightening the drain plug, remember that while it must be securely fastened, there is no need to strain yourself – you will not achieve anything except to risk shearing off the drain plug or damaging the sump. Remember also that it may be you who has to undo it, next time!

Move the container under the oil filter. On 1.3 litre models, this is at the rear left-hand end (seen from the driver's seat) of the engine, and can be unscrewed from above.

On early Maestro 1.6 litre models with the 'R-series' engine, the filter is screwed into the front of the sump, and is reached from below.

On all other 1.6 litre models, the filter is at the rear right-hand end (seen from the driver's seat) of the engine, and can be unscrewed from below. On all 2.0 litre models, the filter is at the right-hand end (seen from the driver's seat) of the engine; on early models it is at the

▲ Unscrewing engine oil filter – early 2.0 litre model

front and can be unscrewed from above or below, while on later models it is at the rear and can be reached from below.

On all models, wipe clean the area around the filter and unscrew it, using a suitable oil filter wrench if necessary to loosen it. Be prepared for the spillage of warm oil. As you remove the filter, check that its sealing ring is removed with it, and does not stick to the engine.

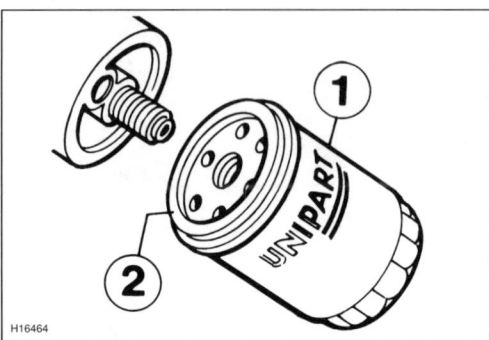

▲ *Engine oil filter renewal*
 1 *Filter*
 2 *Sealing ring – apply smear of clean engine oil before refitting*

Wipe clean the oil filter mounting flange on the engine. Smear a little clean engine oil on the sealing ring of the new oil filter, then screw the filter onto the engine, and tighten it *by hand only*. If no instructions are provided with the filter, tighten it until the sealing ring touches the mounting flange on the engine, then tighten it a further half-turn.

Refill the engine with the correct quantity and grade of oil through the filler on top of the engine. Fill over a period of several minutes, until the level reaches the 'MAX' mark on the dipstick.

Make sure that the oil filler cap is fitted on completion.

Don't race the engine when starting it for the first time after an oil change, as the oil will take a few seconds to circulate. There may be a delay before the oil pressure warning light goes out, as the engine lubrication system fills with oil.

Run the engine, check for leaks from the filter and the drain plug, then stop the engine and check the oil level. Refer to *'Regular checks'* on page 83 for details of how to check the oil level. It is quite likely that a small amount will have to be added, to bring the level up to the 'MAX' mark on the dipstick.

Dispose safely of the old engine oil and filter (refer to *'Safety first!'* on page 98). **Don't** pour it down a drain.

COOLING SYSTEM

Check hoses, clips and joint surfaces

Refer to the *'ENGINE COMPARTMENT'* checks on page 103.

Check, and adjust or renew, the alternator drivebelt

Note: *If the drivebelt requires adjustment, suitable spanners (or sockets) will be required to fit the alternator pivot and adjuster link bolts and nuts.*

The alternator is driven by a V-shaped (plain or toothed) or flat, multi-ribbed drivebelt from the crankshaft pulley, these being found on the right-hand end of the engine (seen from the driver's seat). On 1.3 litre models, and early Maestro 1.6 & 2.0 litre models, the same drivebelt also drives the engine's cooling system water pump.

On 1.3 litre models, and early Maestro 1.6 litre models with the 'R-series' engine, the best access is gained from above, reaching down to the alternator behind the engine. On all other 1.6 litre models and later 2.0 litre models, the best alternator access is from above, in front of the engine. On early 2.0 litre models, jack up the front right-hand side of the car and support it securely on axle stands (refer to *'Jacking and vehicle support'* in *'Safety first!'* on page 98 for details of where to position the jack and stands), then remove the roadwheel (refer to *'Breakdowns'* on page 49) and remove its retaining screws to release the access panel from under the wheel arch; the alternator and drivebelt can then be examined.

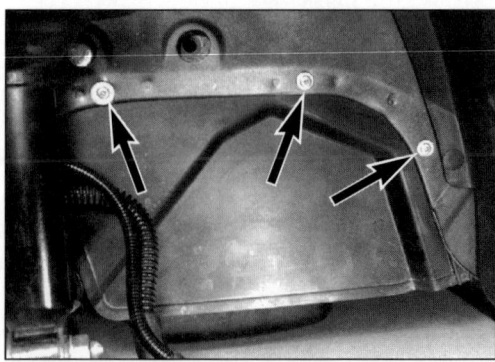

▲ *Early 2.0 litre models – remove screws (arrowed) and withdraw access panel to reach alternator*

Examine the drivebelt for signs of cracking, obvious wear, or contamination, and renew it if necessary. On 2.0 litre models with power-assisted steering, it will be necessary first to remove the pump drivebelt (refer to *'SUSPENSION AND STEERING'* on page 115 for details of how to do this) before the alternator drivebelt can be disturbed.

When a new belt is fitted, the tension should be rechecked after the car has covered 1000 miles (1500 km).

To check the tension of the belt, press down on the belt midway between the pulleys on the belt's top run (where accessible – an alternative is shown for early 2.0 litre models); the deflection of the belt should be 9 mm (0.4 in) under moderate finger pressure.

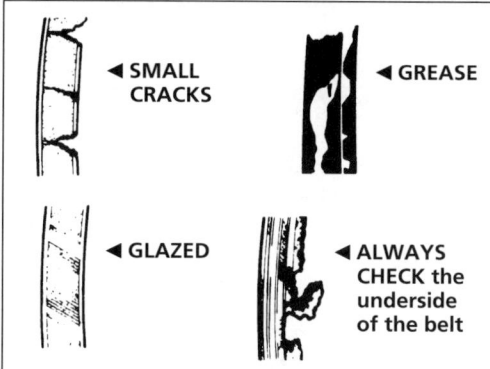

◀ SMALL CRACKS   ◀ GREASE

◀ GLAZED   ◀ ALWAYS CHECK the underside of the belt

▲ *Examine the alternator drivebelt for signs of wear, deterioration or contamination – V-belt*

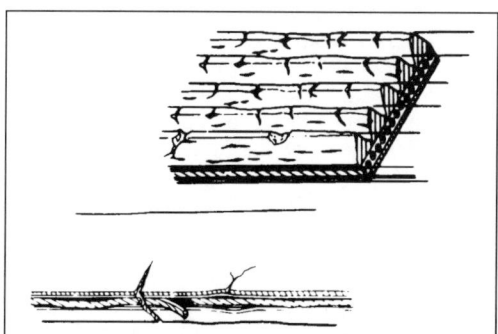

▲ *Examine the alternator drivebelt for signs of wear, deterioration or contamination – flat, multi-ribbed belt*

To renew the belt, slacken all the pivot and adjuster link bolts and nuts, then move the alternator towards the engine to slip the belt off the pulleys.

Fit the new belt, pull the alternator away from the engine until the belt is fairly tight, then tighten the bolts and nuts.

Run the engine (at no more than a fast idle speed) for five minutes.

Tension the belt as described below.

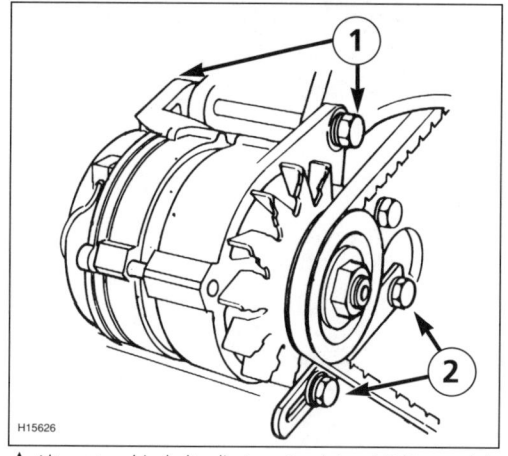

▲ *Alternator drivebelt adjustment points – 1.3 litre models*
1 *Pivot bolt(s) and nut(s)*
2 *Adjuster link bolts (and/or nuts)*

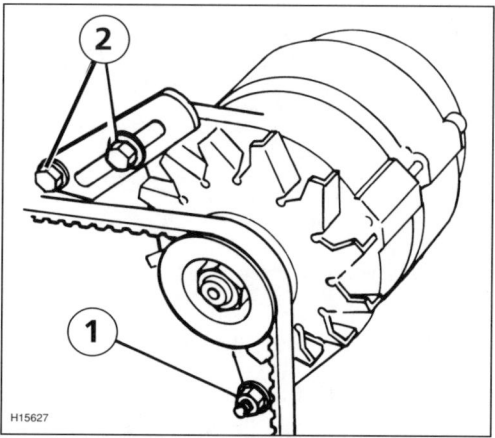

▲ *Alternator drivebelt adjustment points – early Maestro 1.6 litre models ('R-series' engine)*
1 *Pivot bolt(s) and nut(s)*
2 *Adjuster link bolts (and/or nuts)*
Note: *Later 2.0 litre models similar.*

AUSTIN/MG MAESTRO & MONTEGO

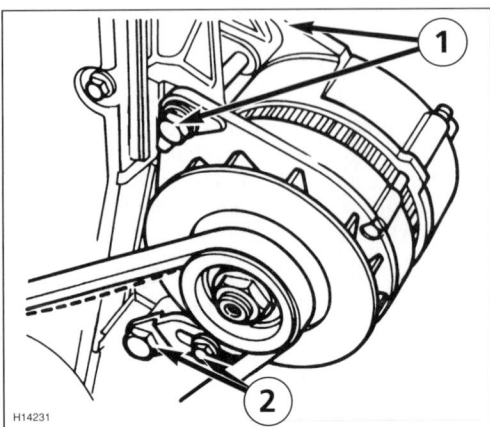

▲ *Alternator drivebelt adjustment points – all other 1.6 litre models*
1 *Pivot bolt(s) and nut(s)*
2 *Adjuster link bolts (and/or nuts)*

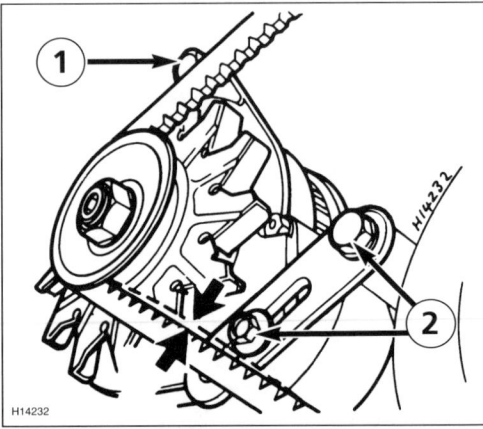

▲ *Alternator drivebelt adjustment points – early 2.0 litre models*
1 *Pivot bolt(s) and nut(s)*
2 *Adjuster link bolts (and/or nuts)*
Note: *Arrows indicate accessible checking point.*

To adjust the tension, slacken the alternator pivot bolts and nuts **(1)** and the adjuster link bolts and nuts **(2)**, then move the alternator towards or away from the engine as necessary to give the correct tension. If leverage is required, exert pressure **only** against the alternator drivebelt end, using a wooden or soft metal lever.

Once the tension is correct, tighten the

adjuster link alternator end fastening, adjuster link engine end fastening, and then the alternator pivot, in that order. Where two alternator pivot bolts and nuts are fitted, tighten the drivebelt end pivot bolt and nut first, then finally the remaining pivot bolt and nut.

FUEL AND EXHAUST SYSTEMS

Check the condition of the fuel pipes and hoses

Refer to the *'ENGINE COMPARTMENT'* checks on page 103, and to the *'CAR UNDERSIDE'* checks on page 118.

Check the condition of the fuel tank, and of the exhaust system

Refer to the *'CAR UNDERSIDE'* checks on page 118.

Top-up the carburettor piston damper

Note: *This task does not apply to the Weber carburettors fitted to MG Maestro 1600 models. For all other (SU-) carburettor-engined models, there is no need to buy a special carburettor oil for this task; engine oil (of the recommended grade and viscosity – refer to* **'Service specifications'** *on page 77) is perfectly adequate.*

The carburettor piston damper is reached by unscrewing the black plastic cap from the top of the carburettor. Also known as the carburettor 'dashpot', it must be topped-up at least at this interval, and should be checked whenever the car's acceleration and/or smooth running at low speeds seems to have suffered.

Note that MG Turbo models have the damper cap secured by a clamp; this must be released to unscrew the damper, and must be securely tightened on refitting the damper.

To top-up the damper, unscrew the cap and carefully pull it, with the plunger, out of the carburettor.

Using an oil can filled with the same type of oil as that recommended for the engine (refer to 'Service specifications'), top the damper up to the top of the hollow piston rod. Don't overfill it, or some may spill over the carburettor, but if you do, don't worry too much; a small surplus will merely be burned by the engine as soon as you start it up.

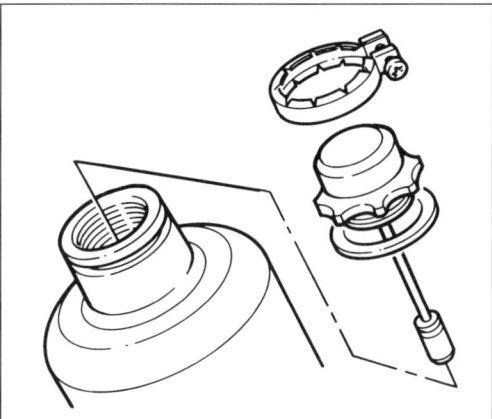

▲ *Carburettor piston damper securing clamp – MG Turbo models*

▲ *Topping-up the carburettor piston damper*

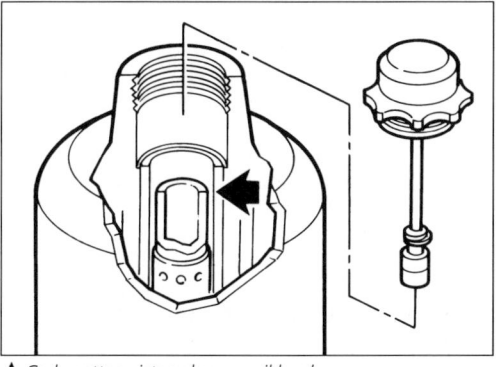

▲ *Carburettor piston damper oil level*

Carefully press the damper plunger back into the piston rod, and slide it down until the cap can be screwed into place (be careful, the plastic threads are easily damaged) and tightened securely; do not use excessive force when refitting the damper or tightening the cap.

Check operation of accelerator and linkage (and manual choke, where fitted)

Remove the air cleaner, if required, to reach the top of the carburettor. Trace the accelerator cable (and choke, where fitted) from the engine compartment bulkhead to the carburettor, checking that there are no signs of external wear or damage, and that the cable is routed in wide, easy loops (with no sharp bends or kinks). Also check that the cable is secured by clips or ties, clear of any other component which will become hot in use (such as the exhaust), or which will move around and might rub against the cable.

Have an assistant sit in the car and operate the accelerator pedal, then the choke control; check that each control moves smoothly and easily through its full travel, and returns equally smoothly and easily to the at-rest position. It is essential that the throttle in particular snaps shut as soon as the pedal is released.

If any sign of wear, damage or faulty operation is found, either have the offending cable renewed, or do it yourself with reference to the Owners Workshop Manual.

When the cables are known to be in good condition, use an oil can to apply a few drops of oil (the same type as recommended for the engine) to the exposed ends of the cables and to their linkages on the carburettor, then refit the air cleaner (where removed).

Renew the air cleaner filter element

1.3 litre models, and early Maestro 1.6 litre models (except MG) with the 'R-series' engine

Unscrew the three securing screws **(A)** and the single screw **(B)**, collecting their washers, and lift the air cleaner, checking that the seal remains in place on the carburettor air intake, and disconnecting the hoses and pipes as necessary.

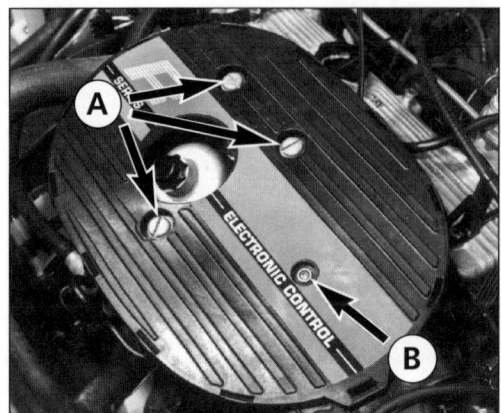

▲ *Remove air cleaner securing (A) and cover (B) screws ...*

▲ *... unclip cover ...*

▲ *... and withdraw filter element – 1.3 litre models and 1.6 litre 'R-series' engine models except MG*

Separate the air cleaner cover from the body by prising it up at each of the cover's retaining clips in turn, working around the cover until it is free. Lift out the old element then, using a clean cloth, wipe out the inside of the cleaner body.

Fit the new element and replace the cover, then press the two together until the cover clips into place and is securely fastened.

Carry out any other carburettor service procedures which require air cleaner removal, then refit the air cleaner assembly, ensuring that the vacuum pipe is not trapped, that the seals are not dislodged, and that all washers are refitted. Tighten the screws securely.

MG Maestro 1600 models

Disconnect the crankcase breather hose from the air cleaner cover. Unscrew the nuts, collecting their washers (3), and release the clips to lift off the cover.

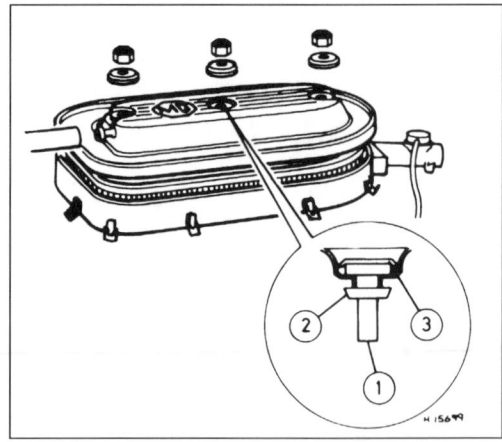

▲ *Air cleaner assembly – MG Maestro 1600 models*
1 Spacer   2 Seal   3 Washer

Lift out the old element then, using a clean cloth, wipe out the inside of the cleaner body.

Fit the new element (it will only fit correctly one way) and replace the cover, ensuring that the spacers (1) and seals (2) are correctly located. Secure the clips, check that the washers (3) are correctly in place, and tighten the nuts securely before reconnecting the crankcase breather hose.

All other 1.6 litre models, and early 2.0 litre models – drum-shaped air cleaner

Disconnecting the vacuum hose or pipe if

▲ *Renewing filter element – drum-shaped air cleaner*

▲ *Removing a distributor cap for cleaning (1, 2, 3 and 4 show HT lead contacts by cylinder numbers, A is carbon centre contact)*

necessary, lift and release the clips securing the air cleaner cover, and lift out the used element.

Using a clean cloth, wipe out the inside of the cleaner body, fit the new element and replace the cover, ensuring that the clips are securely fastened and the hose/pipe (where disconnected) is securely refastened.

Later 2.0 litre models – flat air cleaner

Release the clips securing the air cleaner cover and work the used element out of the cleaner body. Using a clean cloth, wipe out the inside of the cleaner body, fit the new element and replace the cover, ensuring that the clips are securely fastened.

IGNITION SYSTEM

Clean the distributor cap, coil and HT leads

Remove the distributor cap, which may be secured with spring clips or screws.

Thoroughly clean the cap inside and out with a dry lint-free cloth.

Examine the inside of the distributor cap, and if any of the four HT lead contacts inside the cap appear badly burnt or pitted, or if there are any signs of hairline cracks in the plastic, renew the cap. While you are there, check the rotor arm and renew it if it shows similar faults, particularly if its brass contact is burned or worn.

Make sure that the carbon contact in the

centre of the cap is not excessively worn, that it's free to move, and that it protrudes from its holder – if not, renew the cap.

Disconnect the HT lead from the ignition coil, trace it through to the connection on the distributor cap, and remove the lead. Check the contacts on both ends of the lead for corrosion, and clean any away if found; also wipe the lead clean over its entire length. Wipe the ignition coil clean, check the condition of its HT lead contact, and similarly inspect the contact on the distributor cap before refitting the lead.

Now disconnect one of the four remaining HT leads from the distributor cap, trace it through to its spark plug, and remove it. Inspect and clean the lead as described above for the coil lead, and inspect the contact on the distributor cap before refitting the lead. Repeat these operations on the remaining three HT leads, removing, cleaning and refitting one lead at a time, to avoid mixing them up.

If any of the leads is damaged or badly corroded, renew all five leads as a set.

Renew the spark plugs

Note: *Ensure that a set is obtained of the correct type of new spark plug for your particular car. A suitable spark plug spanner, and either a spark plug gap adjustment tool or a set of feeler gauges will be required for this task.*

The spark plugs must be renewed with the engine *cold*.

The spark plugs and their HT leads are numbered 1 to 4, starting from the alternator drivebelt end of the engine – the number 1 spark plug is shown on all the underbonnet component location illustrations in this Chapter. The plugs can be renewed in any order you like, as long as they're all done; for simplicity, it's probably best to work from one end of the engine to the other.

▲ *Checking a spark plug gap using an adjustment tool*

▲ *Adjusting a spark plug gap using the correct tool*

Starting with the number 1 spark plug, pull the HT lead from the plug – pull on the rubber insulator at the end of the lead, not on the lead itself.

Before removing the spark plug, brush any grit from the area around the plug recess to avoid it dropping into the engine as the plug is removed.

Using a proper spark plug spanner (to avoid cracking the ceramic insulation), unscrew the spark plug and remove it from the engine.

Before fitting the new plug, the electrode gap must be set as follows (refer to '*Service specifications*' on page 79 for the correct electrode gap).

Measure the gap between the electrodes using a spark plug gap adjustment tool, or a feeler gauge. The gap is correct when the tool or gauge is a firm sliding fit between the electrodes.

If adjustment is required, carefully bend the **outer** electrode until the correct gap is obtained. **Never** try to bend the centre electrode.

Make sure that the plug threads and the seating area in the cylinder head are clean, and apply a thin smear of graphite grease or copper-based anti-seize compound (available from most good motor accessory shops) to the plug threads. Screw the plug in by hand initially.

Using the spark plug spanner, tighten the plug until initial resistance is felt, then tighten by a further quarter of a turn – no more; **do not** overtighten the plug.

Refit the HT lead, making sure that it is a secure fit over the end of the plug – there should be a soft 'click' as the lead is connected.

Repeat the above procedures for the remaining three spark plugs in turn, removing only one HT lead at a time, to avoid confusion.

On completion, check once more that all four HT leads are securely fitted.

MANUAL GEARBOX

Check the operation of the clutch pedal

Refer to '*ROAD TEST*' on page 118.

Check for oil leaks

Refer to the '*ENGINE COMPARTMENT*' checks on page 103, and to the '*CAR UNDERSIDE*' checks on page 118.

Check the condition of the driveshaft joints and rubber gaiters

Refer to the *'CAR UNDERSIDE'* checks on page 118.

AUTOMATIC TRANSMISSION

Check for fluid leaks

Refer to the *'ENGINE COMPARTMENT'* checks on page 103, and to the *'CAR UNDERSIDE'* checks on page 118.

Check the fluid level

To obtain an accurate reading, the fluid level must be checked with the engine and transmission at normal operating temperature, preferably after a journey of at least 10 miles (15 km). The level can be checked with the engine cold, but it's not advisable to make it a habit – refer to the note below.

Park the car on level ground, select the 'P' position, then apply the handbrake.

With the engine running at its normal idle speed, apply the brake pedal and move the transmission selector lever three times through the full range of positions, then select **N** (1.6 litre models) or **P** (2.0 litre models).

With the engine still idling, pull out the automatic transmission fluid level dipstick (located at the front left-hand side of the engine compartment – when viewed from the driver's seat), and wipe it dry with a clean lint-free rag.

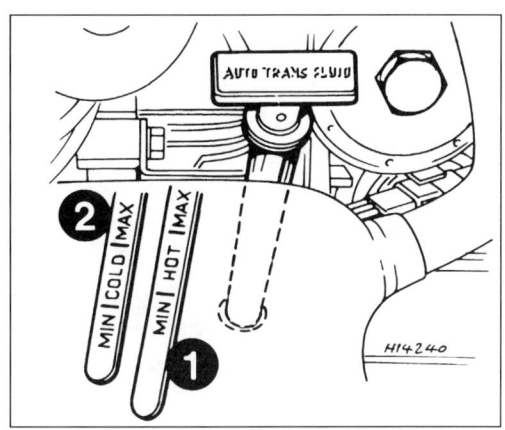

▲ *Automatic transmission fluid level dipstick location and level markings – 2.0 litre models*
1 (Engine) Hot markings – accurate check
2 (Engine) Cold markings – rough guide only

Fully re-insert the dipstick, then withdraw it again and check the fluid level, which should be between the 'O' marks (1.6 litre models) or between the 'MAX' and 'MIN' marks on the 'Hot' side of the dipstick (2.0 litre models). Repeat if necessary to obtain an accurate reading.

If topping-up is necessary, use only the specified fluid type (refer to *'Service specifications'* on page 77). Top-up by pouring the fluid through the dipstick tube. Take care not to overfill – the level **must not** exceed the upper 'O' (or 'MAX') mark; the amount of oil required to raise the level from the lower dipstick mark to the upper mark is approximately 0.4 litre (0.70 pint) on 1.6 litre models, and 0.3 litre (0.53 pint) on 2.0 litre models.

If there's a need for regular topping-up, it's likely that there's a fluid leak, and the problem should be referred to a Rover dealer.

Note: *If, for any reason, the fluid level cannot be checked with the engine/transmission warmed up, on 1.6 litre models simply check that the level is above the dipstick lower 'O' mark when the engine/transmission is cold; if the level is lower than this, top-up only to the lower mark, then start the engine, warm it up and make an accurate check as described above. On 2.0 litre*

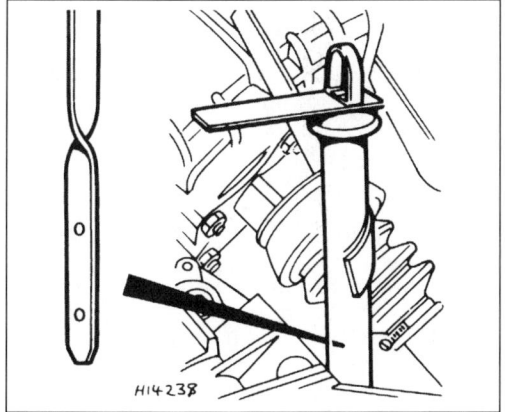

▲ *Automatic transmission fluid level dipstick location and level markings – 1.6 litre models*

models, make the check as described, but using the 'COLD' side of the dipstick, then make an accurate check as described above.

Check the condition of the driveshaft joints and rubber gaiters

Refer to the *'CAR UNDERSIDE'* checks on page 118.

BRAKING SYSTEM

Check the condition of the pipes, flexible hoses and their unions

Refer to the *'ENGINE COMPARTMENT'* checks on page 103 and to the *'CAR UNDERSIDE'* checks on page 118.

Check the operation of the warning lamps fitted

If any of the braking system warning lamps fail to light, either have the bulb renewed by a Rover dealer, or do it yourself – refer to the Owners Workshop Manual for your car (OWM 922 for Maestro 1.3 & 1.6 models, OWM 1066 for Montego 1.3 & 1.6 models, or OWM 1067 for Montego 2.0 models); if the lamp still fails to light, have the circuit checked.

If any of these warning lamps indicate a braking system fault at any time, check as far as you are able whether a fault actually exists, and establish whether or not the car is safe to drive before taking it to a Rover dealer; if you are in any doubt at all, **do not** drive the car – call for assistance.

Check the condition and security of the brake servo vacuum hose

Refer to the *'ENGINE COMPARTMENT'* checks on page 103.

Check the operation of the brakes

Check that the handbrake operates correctly, without excessive movement of the lever, and also that normal pedal pressure is present before taking the car out on the road. Refer to the *'ROAD TEST'* on page 119.

SUSPENSION AND STEERING

Check the tightness of the roadwheel nuts, and check the wheels for damage

Remove the wheel trim, where fitted (refer to

'Breakdowns' on page 51 for details); slacken each nut through one-quarter of a turn, then tighten it again securely. If a torque wrench is available, have the nuts slackened and then retightened to the correct pressure; this is especially important if the car has light alloy wheels fitted.

If the nuts are very tight or stubborn, now is a good time to remove the wheel completely, so that it can be cleaned inside and out and the mating surfaces lubricated (refer to 'Changing a wheel' on page 54); this will minimise the trouble involved in dealing with any future roadside punctures.

Check around both the outside and inside edges of each roadwheel for signs of damage or distortion, and make sure that any wheel balance weights are secure, with no obvious signs that any are missing.

If there is any evidence of damage to a wheel, seek advice immediately, and if necessary renew the wheel.

If a wheel balance weight is obviously missing, some vibration can be expected whilst driving, and the wheel should be re-balanced.

Check the power-assisted steering fluid level

On 1.6 litre models, the power-assisted steering pump is located at the rear right-hand (seen from the driver's seat) end of the engine, between the braking system master cylinder/fluid reservoir and the carburettor. On 2.0 litre models, the pump is at the front right-hand (seen from the driver's seat) end of the engine.

The pump has the power-assisted steering fluid reservoir integral in its upper end.

The fluid level must be checked when the system is *cold,* and with the engine stopped; it is essential for an accurate reading that the steering should not have been turned since the engine was switched off.

To check the fluid level, first wipe clean the top of the pump to prevent any risk of dirt falling in when the filler cap is removed, then unscrew the cap (anti-clockwise) and withdraw it; wipe the dipstick dry with a clean lint-free rag. Fully refit the filler cap, then withdraw it again and check the fluid level, which should be between the 'MAX' and 'MIN' marks on the dipstick.

If topping-up is necessary, use only the

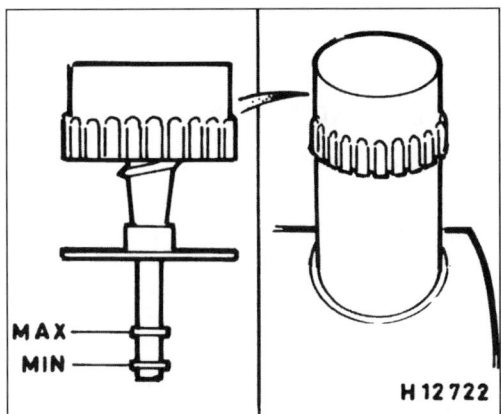

▲ *Power-assisted steering fluid level dipstick location and level markings*

specified fluid type (refer to *'Service specifications'* on page 77). Take care not to overfill – the level **must not** exceed the 'MAX' mark. Ensure that the filler cap is securely refitted on completion.

If there's a need for regular topping-up, it's likely that there's a fluid leak, and the problem should be referred to a Rover dealer; do not use the car if the fluid is below the dipstick 'MIN' mark. If the loss is serious, and the car is being driven to a Rover dealer for attention, **do not overfill** the reservoir to compensate for fluid loss while you are in motion; in such cases, the pump drivebelt should be removed to protect the pump from running with no fluid, but remember that the steering effort required would then be far greater than normal.

Check, and adjust or renew, the power-assisted steering pump drivebelt

Note: *If the drivebelt requires adjustment, suitable spanners (or sockets) will be required for the pump pivot and adjustment bolts and nuts.*

The pump is driven by a V-shaped (plain or toothed) drivebelt from the camshaft (1.6 litre models), or from the water pump (2.0 litre models) pulley, these being found on the right-hand end of the engine (seen from the driver's seat).

To examine the drivebelt on 1.6 litre models, unclip any hoses, then unscrew the single

bolt **(3)** and withdraw the drivebelt cover.

Examine the drivebelt for signs of cracking, obvious wear, or contamination, and renew it if necessary (refer to the alternator drivebelt check, under *'COOLING SYSTEM'* on page 107 for details).

To renew the belt, slacken all the pivot and adjustment bolts and nuts (also the high-pressure pipe nut **(4)** on 2.0 litre models), then move the pump towards the engine to slip the belt off the pulleys; refer to the note below concerning handling the pump. Fit the new belt, pull the pump (carefully) away from the engine until the belt is fairly tight, tighten the bolts and nuts, then tension the belt as described below.

To check the tension of the belt, press down on the belt midway between the pulleys on the belt's top run; the deflection of the belt should be 9 mm (0.4 in) under moderate finger pressure.

To adjust the tension, slacken the pump pivot bolts (and nut) **(1)** and the adjustment bolts (and nut) **(2)**; on 2.0 litre models, hold the pump nut **(3)** with one open-ended spanner and just slacken the high-pressure pipe nut **(4)** with a second open-ended spanner, until the pipe can be moved with no more than a slight trace of fluid appearing. Move the pump

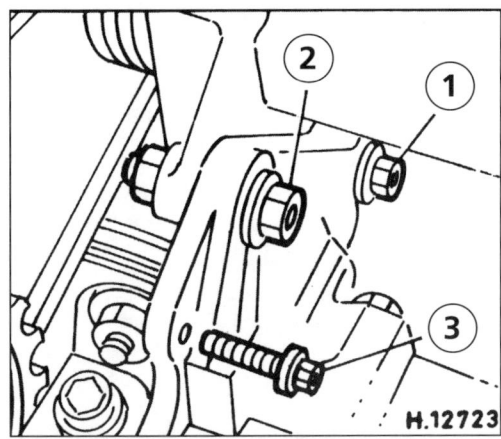

▲ *Power-assisted steering pump drivebelt adjustment points – 1.6 litre models*
1 *Pivot bolt and nut*
2 *Adjustment bolt and nut*
3 *Belt cover securing bolt*

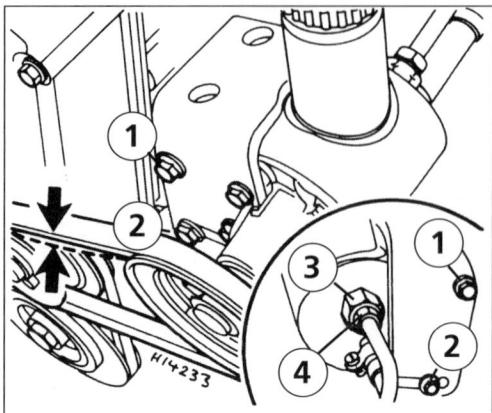

▲ *Power-assisted steering pump drivebelt adjustment points – 2.0 litre models*
1 *Pivot bolts*
2 *Adjustment bolts*
3 *Pump nut*
4 *High-pressure pipe nut*
Arrows show checking point for belt tension

towards or away from the engine as necessary to give the correct tension.

Note: *If leverage is required, exert pressure* **only** *against the pump body or mounting bracket using a wooden or soft metal lever;* **do not** *lever against, or apply other pressure to, the reservoir.*

Once the tension is correct, tighten the adjustment bolts (and nut), then the pivot bolts (and nut). On 2.0 litre models, hold the pump nut while tightening the high-pressure pipe nut; if more than a slight amount of fluid was lost, check the fluid level and top-up if necessary. On 1.6 litre models, refit the belt cover and secure the hoses in their clips.

Check the front and rear suspension struts for fluid leaks

Jacking up each corner of the car in turn and supporting it securely on axle stands (refer to *'Jacking and vehicle support'* in *'Safety first!'* on page 98 for details of where to position the jack and stands), check around the body of each strut for signs of dampness, which may be due to fluid leakage.

If any fluid is noticed, seek advice, and if necessary have the strut renewed; it is usually

considered good practice to have such components renewed in matched pairs, as axle sets.

INTERIOR AND BODYWORK

Check the condition and security of the seats and seat belts

If any of the seat belt webbing is damaged, frayed or worn, the relevant seat belt(s) must be renewed.

Check that all inertia reel seat belts can be pulled smoothly from their retractors, and that they return smoothly, unassisted.

Check that the locking mechanism works correctly by sharply tugging the seat belt webbing.

If a seat belt retractor or locking mechanism appears to be faulty, the relevant seat belt should be renewed.

Check that all seat and seat belt mountings are in good condition and securely fastened.

Lubricate all hinges, door locks, and the bonnet release mechanism

Use general-purpose light oil for the door locks, and a suitable general-purpose grease for the hinges and bonnet release mechanism. **Do not** attempt to lubricate the steering lock; if it is stiff, seek the advice of a Rover dealer.

Don't use an excess of lubricant, or it may find its way onto other components, or onto the clothing of the car's occupants!

Check the condition of the underseal

Refer to the *'Spring – UNDERBODY'* on page 122.

Inspect the paintwork and bodywork

Touch up any minor stone chips, and have any corroded areas repaired before the damage becomes serious.

For details of how to treat minor paint scratches, refer to *'Bodywork and interior care'* on page 126.

At regular intervals, use a length of stiff wire to probe clear the drainage points located in the sunroof aperture (where fitted), along the lower edge of the doors and tailgate/boot lid and around the sills and rear of the car.

▲ *Body drain points – Maestro*

▲ *Body drain points – Maestro*

ELECTRICAL SYSTEM

Check the condition of all accessible wiring connectors, the wiring harness and its clips

While you are working around the car, carrying out the other tasks in this service, check that all visible wiring is in sound condition. The connectors should be securely fastened, and the wiring secured out of harm's way (ie, clear of any hot or moving components) by the clips provided.

Check, and if necessary clean, the battery terminals

Make sure that the terminals are secure, and clean any corrosion from the metal. Be sure to wear eye protection for this, as the white-coloured deposits are harmful. To prevent corrosion, the terminals can be coated with petroleum jelly (don't use ordinary grease). It's worthwhile examining the battery tray at the same time, and if the same white-coloured deposits are present, clean them using a brush (or wash them away with plenty of hot water, being careful to avoid splashes).

Check the headlamp alignment, and have adjusted if necessary

Refer to the Owners Workshop Manual, but note that accurate adjustment is a job for a Rover dealer.

CAR UNDERSIDE

Check all hoses and pipes, and the surfaces of all components for signs of fluid leakage, corrosion, damage or deterioration

Refer to the *'ENGINE COMPARTMENT'* checks on page 103.

In addition, check the exhaust system and the fuel tank for any sign of damage or serious corrosion (light surface rust is to be expected).

Check the exhaust mountings to make sure that they're secure.

Check the visible suspension and steering components for obvious signs of damage or wear.

Check the driveshafts for obvious signs of distortion or damage.

Pay particular attention to the driveshaft and steering gear rubber gaiters, which should be renewed if they're split, or if there are any obvious signs of lubricant leakage.

▲ *Check the driveshaft and steering gear rubber gaiters for splits (arrowed)*

ROAD TEST

Check the operation of all instruments and electrical equipment

Make sure that all instruments read correctly, and switch on all electrical equipment in turn to check that it functions properly.

Check for any abnormalities in the steering, suspension, handling or road feel

Drive the car and check that there are no unusual vibrations or noises.

Check that the steering feels positive, with no excessive 'sloppiness' or roughness, and check for any suspension noises when cornering and driving over bumps.

Check the performance of the engine, clutch (where applicable), gearbox and driveshafts

Listen for any unusual noises from the engine, clutch and gearbox.

Make sure that the engine runs smoothly when idling, and that there's no hesitation when accelerating.

Check that, where applicable, the clutch action is smooth and progressive, that the drive is taken up smoothly, and that the pedal travel is not excessive. Also listen for any noises when the clutch pedal is depressed.

On manual gearbox models, check that all gears can be engaged smoothly, without noise, and that the gear lever action is not abnormally vague or 'notchy'.

On automatic transmission models, make sure that all gearchanges occur smoothly, without snatching, and without an increase in engine speed between changes. Check that all the gear positions can be selected with the car at rest. If any problems are found, they should be referred to a Rover dealer.

Listen for a metallic clicking sound from the front of the car, as the car is driven slowly in a circle with the steering on full lock. Carry out this check in both directions. If a clicking noise is heard, this indicates wear in a driveshaft joint, in which case seek advice, and renew the joint if necessary.

Check the operation and performance of the braking system

Make sure that the car does not pull to one side when braking, and that the wheels do not lock prematurely when braking hard.

Check that there's no vibration through the steering when braking.

Check that the handbrake operates correctly, without excessive movement of the lever, and that it holds the car stationary on a slope.

ADDITIONAL TASKS

Note that as well as the tasks described in the preceding paragraphs, the additional tasks given in the *'Service schedule'* chart on page 90 should be carried out every 12 000 miles (20 000 km) or 1 year – whichever comes first. These tasks require more detailed explanation, or the use of special tools, and are considered beyond the scope of this Handbook. For details of these additional tasks, refer to the Owners Workshop Manual.

Every 24 000 miles (40 000 km) or 2 years – whichever comes first

---

In addition to all the items in the *'12 000 miles (20 000 km)'* service on page 103, carry out the following.

ENGINE

Renew the engine oil filler cap – 1.3 litre models

The oil filler cap at the top of the engine contains an air filter; it is not always possible to clean this effectively, so to prevent oil leaks and to ensure that the engine runs smoothly and cleanly at low speeds, the cap **must** be renewed at this interval. Remove the cap, pull it off the rocker cover and fit the new cap.

Renew the engine breather filter – Maestro 1.6 litre models with 'R-series' engine

The engine breather is located at the rear left-hand (seen from the driver's seat) end of the cylinder block. It contains an air filter which it is not always possible to clean effectively, so to prevent oil leaks and to ensure that the engine runs smoothly and cleanly at low speeds, the filter **must** be renewed at this interval.

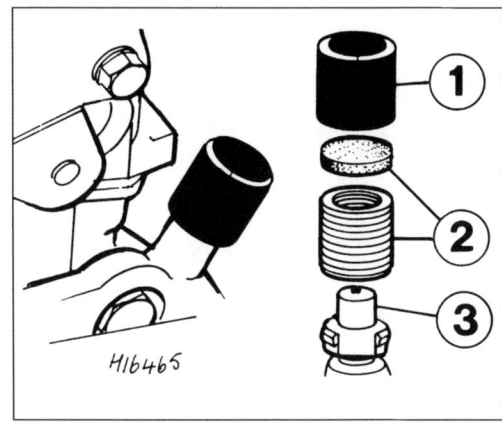

▲ *Engine breather filter assembly – early Maestro 1.6 litre models ('R-series' engine)*
1 *Cover*
2 *Filter and felt pad*
3 *Adaptor*

Remove the breather cover **(1)** by pressing it sideways to release it from the adaptor **(3)**, extract the filter and felt pad **(2)** and unscrew the adaptor. Wipe the housing clean and screw in the new filter, then reassemble and refit the breather.

Clean the engine breather filter – all other 1.6 litre models and 2.0 litre models

The oil filler cap contains an air filter which must be cleaned at this interval, to prevent oil leaks and to ensure that the engine runs smoothly and cleanly at low speeds.

Remove the cap, disconnecting any hoses, and swill it in a bath of solvent until it is clean. Do not use petrol as a cleaning agent, because of its attendant health and safety risks (refer to *'Safety first!'* on page 96). If a proprietary cleaner is not available, a low-flash point solvent such as paraffin or white spirit may be used, provided that great care is taken to prevent any risk of fire, and that you wash your hands carefully afterwards.

If the cap is heavily fouled with oil or 'mayonnaise' it must be renewed; if the deposits inside the engine and oil filler tube seem very heavy, seek expert advice.

When the filter is quite clean, blow it dry using compressed air (or leave it to dry naturally, which will take much longer) and refit it to the engine, connecting any disturbed hoses.

COOLING SYSTEM

Renew the coolant

Note: *The following items will be required for this task:*
- *A suitable quantity of coolant solution (made up from approximately one-third antifreeze and two-thirds clean water)*
- *Suitable container to catch the old coolant as it drains (make sure that it is large enough for the volume of coolant in the system)*
- *Suitable spanner or socket to fit the cylinder block drain plug (where applicable)*
- *A suitable screwdriver to fit the radiator hose clip(s)*

Draining the system

The coolant should be drained with the engine cold.

Set the heater temperature control inside the car to the fully-hot position.

Remove the cooling system filler cap, – refer to the *'Checking coolant level'* on page 84.

Position the container beneath the radiator bottom hose connection, then slacken the hose clips, release the hose and allow the coolant to drain.

A cylinder block drain plug may also be fitted; on early 1.3 litre models it is on the front left-hand end (seen from the driver's seat), while on 'R-series' 1.6 litre engines it is on the rear left-hand end. Where a drain plug is fitted, reposition the container, then unscrew the plug and allow the remaining coolant in the cylinder block to drain into the container.

If rust or sludge is evident in the coolant which has been drained, the cooling system should be flushed as described below, to remove any further contamination from the engine and radiator.

If the coolant drained is clean, then the cooling system can be refilled as described later in this Section.

Flushing the system

Insert the end of a garden hose into the expansion tank, and flush the system through using cold water.

Continue flushing until clean water flows from the disconnected radiator bottom hose (and cylinder block drain plug, where fitted).

While this procedure should serve to keep the system clear for all normal purposes, a more thorough flushing may be required if cooling system maintenance has been neglected. This will require the disconnection of several hoses, possibly the removal of the thermostat, and may involve reverse-flushing; refer to the Owners Workshop Manual.

If it seems that the system is particularly badly contaminated, and flushing does not appear to be solving the problem, seek advice, as it may be possible to clean the radiator using a proprietary cleaning compound.

Refilling the system

Reconnect the radiator bottom hose and tighten the clips, refit and tighten the cylinder block drain plug (where fitted).

A solution of two-thirds clean water and one-third antifreeze should be used to fill the system all year round; a stronger (50% antifreeze to 50% water) solution can be used in exceptionally cold conditions. Seek the advice of a local Rover dealer if in doubt. It's important to note that because of the different types of metals used in the engine, it's vital to use antifreeze with suitable anti-corrosion additives all year round. **Never** use plain water to fill the whole system. If there's any doubt about the strength of the coolant mixture, it's better to add too much antifreeze than not enough.

Fill the system slowly through the expansion tank until it is full. Wait a few moments for trapped air to escape, squeezing the radiator top hose to help, then add more coolant. Repeat this procedure until the level does not drop any more.

Start the engine and run it (at idle speed) for three minutes, then stop it and check the coolant level; refer to *'Regular checks'* on page 84. Refit the filler cap.

Start the engine again and run it to normal operating temperature, then check that the heater works and switch off the engine.

Once the engine has cooled, squeeze the hoses and check for air locks (it should be possible to feel coolant in all the hoses).

If the heater failed to give warm air, there may be an air lock in the heater, in which case the air can be forced out by squeezing the heater hoses (which run to the bulkhead at the back of the engine compartment).

If necessary, air can be removed from a hose by loosening a hose clip, and squeezing the hose until all the air is forced out (obviously some coolant will be forced out too).

Recheck the coolant level and top-up as necessary.

AUTOMATIC TRANSMISSION

Check the operation of the parking pawl

With the car standing on a level surface, switch off the engine, select position **P** and release the handbrake.

Try to push the car backwards and forwards as hard as you can; there may be a small amount of rocking, due to play in the driveshafts and in the transmission itself, but the car should not be able to move at all.

If you can move the car, immediately seek the advice of a Rover dealer and, until the fault is repaired, be very careful to apply the handbrake firmly when parking.

ADDITIONAL TASKS

Note that as well as the tasks described in the preceding paragraphs, the additional tasks given in the 'Service schedule' chart on page 90 should be carried out every 24 000 miles (40 000 km) or 2 years – whichever comes first. These tasks require more detailed explanation, or the use of special tools, and are considered beyond the scope of this Handbook. For details of these additional tasks, refer to the Owners Workshop Manual.

SEASONAL SERVICING

If you carry out all the procedures described in the previous Section, at the recommended mileage or time intervals, then you'll have gone a long way towards getting the best out of your car in terms of both performance and long life. In spite of this, there are always other areas, not dealt with in regular servicing, where

neglect can spell trouble.

A little extra time spent on your car at the beginning and end of every Winter will be well worthwhile in terms of peace of mind and prevention of trouble. The suggested tasks which follow have therefore been divided into Spring and Autumn Sections – but it's always a good idea to do them more frequently if you feel able.

Autumn

COOLING SYSTEM

Check the radiator and all hoses for signs of deterioration or damage

Refer to 'ENGINE COMPARTMENT' in the '12 000 miles (20 000 km)' service Section on page 103.

Check the strength of the coolant solution

This can be done using a coolant hydrometer, which should be available from most motor accessory shops and motor factors.

The hydrometer will give an indication of the amount of antifreeze in the system, and should be used exactly in accordance with the manufacturer's instructions.

The strength of the solution should never be allowed to drop below 30% antifreeze, by volume.

If there is any doubt about the strength of the coolant solution, the system should be drained and refilled as described in the '24 000 miles (40 000 km)' service Section on page 119.

ELECTRICAL SYSTEM

Where applicable, check the battery electrolyte level

Refer to 'Regular checks' on page 87.

Check, and if necessary clean, the battery terminals

Refer to 'ELECTRICAL SYSTEM' in the '12 000 miles (20 000 km)' service Section on page 117.

Check the condition and adjustment of the alternator/water pump drivebelt

Refer to the procedure given in the '12 000 miles (20 000 km)' service Section on page 106.

Check the operation of all lights, electrical equipment and accessories

Refer to 'Fault finding' on page 139 if any problems are discovered.

If any of the bulbs need to be renewed, refer to 'Bulb, fuse and relay renewal' on page 127.

Check the washer fluid level, and the wipers and washers

Refer to 'Regular checks' on page 87.

TYRES

Check tread depth and condition

Refer to 'Regular checks' on page 85. Remember that you may be driving in slippery conditions during the Winter.

BODYWORK

Thoroughly clean the car

Wash the car, and then polish it thoroughly to help protect the paint during the Winter.

Spring

UNDERBODY

Thoroughly clean car underside

The best time to clean the underside of the car is after the car has been driven in wet conditions, when the accumulated dirt will be softened up.

To clean the car, first of all the car must be jacked up as high as possible (making sure that it is safely supported – refer to 'Safety first!' on page 98).

Gather together a quantity of paraffin, or water-soluble solvent, a stiff-bristle brush, a scraper, and a garden hose.

With the car jacked up and safely supported, get underneath the car, and cover the brake components at each wheel with polythene bags to stop dirt and water getting into them.

Loosen any encrusted dirt, scraping or brushing it away – the paraffin or solvent can be used where there's oil contamination. Pay particular attention to the wheel arches. Take care not to remove the underseal (a waxy substance applied to the car underbody to protect against corrosion).

When all the dirt is loosened, a wash down with the hose will remove the remaining dirt and mud.

You can now check for signs of damage to the underseal. If there's any sign of the underseal breaking away, patch it up by spraying or brushing on a suitable underseal wax (available from motor accessory shops and motor factors). Make sure the area is clean and dry before applying the wax.

Take the opportunity to check for signs of rusting. Likely places are the body sills and floor panels.

If rust is found, seek advice, and have the affected area repaired before the problem gets too bad.

On completion, lower the car to the ground.

BODYWORK

Thoroughly check and clean the surfaces of the bodywork

Give the car a thorough wash, and check for stone chips and rust spots. For details of how to treat these minor paint problems, refer to 'Bodywork and interior care' on page 126.

After any repairs to the paint have been carried out (not before, otherwise the paint will not stick), give the car a polish.

TOOLS

WHAT TO BUY

If you're intending to carry out your own servicing, you'll need to obtain a few basic tools. Although at first sight you may think that tools seem expensive, once you've bought them they should last a lifetime if you look after them properly.

The tools supplied with the car will enable you to change a wheel, and not much more. The absolute minimum tool kit you'll need to carry out any maintenance or servicing will be a range of suitable spanners, two screwdrivers (one for crosshead or 'Phillips' type screws), and a pair of pliers. With a bit of ingenuity, these items should enable you to complete the more basic routine servicing jobs, but they won't allow you to do much else.

When buying tools, it's important to bear in mind the quality. You don't need to buy the most expensive tools available, but generally you get what you pay for. Cheap tools may prove to be a false economy, as they're unlikely to last as long as better quality alternatives.

It's very difficult to lay down hard and fast rules on exactly what you're going to need, but the following list should be helpful in building up a good tool kit. Combination spanners (ring one end, open-ended the other) are recommended because, although more expensive than double open-ended ones, they give the advantages of both types.

- *Combination spanners to cover a reasonable range of sizes (from say 7 mm and 5/16 in upwards)*
- *Adjustable spanner*
- *Spark plug spanner (with rubber insert)*
- *Spark plug gap adjustment tool*
- *Set of feeler gauges*
- *Screwdriver (Plain) – 100 mm long blade x 6 mm diameter (approx)*
- *Screwdriver (Crosshead) – 100 mm long blade x 6 mm diameter (approx)*
- *Oil filter wrench*
- *Pliers*
- *Tyre pump*
- *Tyre pressure gauge*
- *A suitable key (Torx or Allen type, depending on model) to fit the gearbox oil filler plug (later models only)*
- *Oil can*
- *Funnel (medium size)*
- *Stiff brush (for general cleaning jobs)*
- *Tool box*
- *Hydraulic jack (ideally a trolley jack)*
- *Pair of axle stands*
- *Suitable containers for draining engine oil and coolant*
- *Inspection lamp*

This is by no means a comprehensive list of tools, and you'll probably want to gradually add to your tool box as you discover the need for other tools, especially if you decide to tackle some of the more advanced servicing jobs described in the **Owners Workshop Manual** for your particular car.

CARE OF YOUR TOOLS

Having bought a reasonable set of tools, it's worth taking the trouble to look after them. After use, always wipe off any dirt or grease using a clean, dry cloth before putting them away. Never leave tools lying around after they've been used – a tool rack, or better still a proper toolbox will prove the best way of keeping everything up together. Rags can be wrapped around loose tools to prevent them from rattling if you're going to keep them in the boot of the car.

Feeler gauges should be wiped with an oily cloth from time to time, to leave a thin coating of oil on the metal which will prevent corrosion. Screwdriver blades inevitably lose their keen edges, and a little occasional attention with a file or an oilstone will keep them in good condition.

AUSTIN/MG MAESTRO & MONTEGO

It's well worth spending the time and effort necessary to look after your car's bodywork and interior. Cleaning your car regularly will not only improve its appearance, it will also protect the bodywork against the elements and the grime encountered in everyday driving. It's worth bearing in mind that if your car looks clean and tidy, it will also be worth more money when you come to sell it or trade it in for a newer model.

This Section will help you to keep your car in 'showroom' condition. If you wish to tackle repairs to more serious bodywork damage or corrosion, comprehensive details can be found in our **'Car Bodywork Repair Manual'**.

CLEANING THE INTERIOR

It's a good idea to clean the interior of the car first, before cleaning the exterior, as this will avoid spreading the dirt from inside over the bodywork.

Start by removing all the loose odds and ends from inside the car, not forgetting the ashtrays, glovebox and any interior pockets. Take out any loose mats or carpets, which should be shaken and brushed, and if possible vacuum-cleaned.

The inside of the car can be cleaned with a brush and dustpan, or preferably a vacuum-cleaner. If you can't use your household vacuum-cleaner, it may be worth investing in one of the small 12-volt hand vacuum-cleaners which can be powered from the car battery. If the carpets are very dirty, use a suitable proprietary cleaner with a brush to remove the ingrained dirt. Ideally, it's best to remove the carpets for cleaning, but this is an involved task in most modern cars, and it will probably be easier to leave them in place.

The facia, door trim, seats, and any other items which require attention can now be wiped over with a cloth soaked in warm water containing a little washing up liquid. If the trim is particularly dirty, use one of the proprietary cleaners available from motor accessory shops – a number of different types are available, suitable for vinyl, cloth or leather upholstery, etc, as required. An old nail brush or tooth brush will help to remove any ingrained dirt. On completion, wipe the surfaces dry using a lint-free cloth, and leave the windows open to speed up drying.

The inside of the windscreen and windows should be cleaned using a proprietary glass cleaner, or methylated spirits. Be careful about using certain household products such as washing-up liquid which may leave a smeary film. Finish off by wiping with a clean, dry paper tissue.

Don't forget to clean the boot.

Check for any nicks or tears in the interior trim, seats, headlining, etc. Various repair kits are available from motor accessory shops to suit most types of trim, but if the damage is very serious, the relevant trim panel will probably have to be renewed.

CLEANING THE BODYWORK

Ideally, the car should be washed every week, either by hand (preferably using a hosepipe), or by using a local car-wash. If you're washing the car by hand, use plenty of water to loosen the dirt and dust, and if possible use a suitable car shampoo or wax additive in the water. Any stubborn dirt such as road tar or bird droppings can be removed using methylated spirit or preferably one of the special proprietary cleaning solutions (in which case make sure that the product is suitable for use on car paint) – in either case, wash the affected area down with plenty of water after removing the dirt. **Never** just wipe over a dirty car, as this will scratch the paint very effectively.

AUSTIN/MG MAESTRO & MONTEGO

Two or three times a year, a good silicone or wax polish can be used on the paintwork. It's important to wash the car and remove all stubborn dirt, tar, etc, before using polish, as the polish will effectively seal over the top of any dirt, making it extremely difficult to remove in the future. Good polishes actually form a protective coating over the paint finish, which should help to make future accumulated dirt easier to remove. Always follow the manufacturer's recommendations closely when using polish. Try to avoid getting polish on any unpainted plastic or rubber body parts such as bumpers and spoilers, as it tends to leave a stain when dry. Chrome parts are best cleaned with a special chrome cleaner, as ordinary metal polish will wear away the finish.

Plastic or rubber body parts such as bumpers and spoilers should be cleaned using a suitable proprietary cleaner. A number of different types are available, including colour restorers, and special solvents to remove any stray polish which may have crept onto the plastic when polishing the paintwork. Make absolutely sure that any solvents or cleaners used are suitable, as certain products may attack plastic and/or rubber.

If the paint is beginning to lose its gloss or colour, and ordinary polishing doesn't seem to solve the problem, it's worth considering the use of a polish with a mild 'cutting' action to remove what is in effect a surface layer of 'dead' paint. In this case, follow the manufacturer's instructions, and don't polish too vigorously, or you might remove more paint than you intended!

DEALING WITH SCRATCHES

With superficial scratches which don't penetrate down to the metal, repair can be very simple.

Very light scratches can be polished out by carefully using a suitable 'cutting' polish. Follow the manufacturer's instructions, and take care not to remove too much of the surrounding paint.

If the scratch cannot be polished out, touch-up paint will be required. Touch-up paint is usually available in the form of a touch-up stick from the car manufacturer's dealers, or in various forms from motor accessory shops. You will probably have to quote the year and model of the car in order to make sure that you obtain the correct matching colour, and in some cases you may have to quote a paint reference number which will usually appear on the car's Vehicle Identification Number (VIN) plate under the bonnet (refer to 'Servicing' on page 103).

Lightly rub the area of the scratch with a very fine cutting paste to remove loose paint from the scratch and to clear the surrounding bodywork of polish.

Rinse the area with clean water.

Apply suitable touch-up paint to the scratch using a fine paint brush, and continue to apply fine layers of paint until the surface of the paint in the scratch is level with the surrounding paintwork.

Allow the new paint at least two weeks to harden, then blend it into the surrounding paintwork by rubbing the scratch area with a paintwork renovator or a very fine cutting paste.

If the paint finish requires a lacquer coat (in which case the lacquer will be supplied with the touch-up paint), it should now be applied to the newly painted area and allowed to dry in accordance with the manufacturer's instructions.

Finally, apply a suitable wax polish to the affected area for added protection.

BULBS

A defective exterior lamp can be not only dangerous, but also illegal. Carrying spare bulbs will enable you to renew blown ones as they occur. A failed interior light bulb may be just a nuisance, but a faulty exterior lamp could be a life-or-death matter.

Before assuming that a bulb has failed, check for corrosion on the bulbholder and wiring connections, particularly around the rear lamp assemblies.

● Note that as a safety precaution, the battery earth (black) lead should always be disconnected before renewing a bulb.
● Remember to reconnect the lead on completion.

Bulb ratings

Always make sure that when a bulb is renewed, a new bulb of the correct rating is used.

All the bulbs are of the 12-volt type, and their ratings are as follows:

Bulb	Rating [watts]
Headlamp	60/55
Front sidelamp – bayonet fitting	4
Front sidelamp – capless/wedge-type fitting	5
Front and rear direction indicator lamps	21
Front direction indicator side repeater lamp	5
Rear brake/tail lamp	21/5
Rear sidelamp	5
Reversing lamp	21
Rear foglamp	21
Rear number plate lamp – bayonet fitting	4
Rear number plate lamp – capless/wedge-type fitting	5
Rear number plate lamp – festoon type	5
Courtesy lamp	10
Luggage compartment/boot lamp	10
Glovebox lamp	5
Footwell lamp	5
Cigar lighter illumination bulb	0.36
Automatic transmission selector illumination bulb	3
Hazard warning switch (steering column-mounted) bulb	1.2

BULBS, FUSES & RELAYS

Exterior light bulbs

HEADLAMP

Open the bonnet; the headlamp bulb is reached from within the engine compartment.

On early Maestro models, twist the plastic cover anti-clockwise to release it from the back of the headlamp, then unplug the wiring connector from the back of the headlamp bulb.

On all other models, unplug the wiring connector and pull off the rubber cover from the back of the headlamp.

On all models, release the spring clip arms from the reflector and withdraw the bulb.

Fit the new bulb, taking care not to touch it with your fingers (hold the bulb using a clean

▲ Pull wiring connector from back of headlamp ...

▲ Twist headlamp rear cover anti-clockwise ...

▲ ... then peel off rubber cover to reach bulb – all other models

▲ ... and unplug wiring to reach bulb – early Maestro models

▲ Releasing spring clip arm from headlamp reflector

▲ *Withdrawing headlamp bulb*

cloth or paper tissue); if the glass on a new bulb is inadvertently touched, wipe it with a cloth moistened with methylated spirit. The bulb's three locating tabs will only fit the correct way in the reflector slots. Fit the clip to the reflector and bulb flange as shown.

On early Maestro models, ensure that the wiring is reconnected, and that the plastic cover is securely fastened with the drain hole at the bottom. On all other models, ensure that the rubber cover is securely fitted to the back of the headlamp so that its drain hole is at the bottom (where applicable, its 'TOP' marking will be uppermost) before reconnecting the wiring.

FRONT SIDELAMP

Open the bonnet, disconnect the headlamp wiring and remove the cover from the back of the headlamp, then pull the sidelamp bulbholder from the headlamp reflector.

On some models, the bulb is a bayonet fitting; remove it by pressing it in and twisting it anti-clockwise. On others, the bulb is a capless/wedge type, and is removed by pulling it out of the holder.

Fit the new bulb using a reversal of the removal procedure.

FRONT DIRECTION INDICATOR LAMP

Open the bonnet. Where the rear of the direction indicator lamp cannot be reached directly from within the engine compartment, next to the headlamp, pull the wire ring to release the lamp's retaining spring clip, and withdraw the lamp from the front of the car.

▲ *Pull wire ring to release front direction indicator lamp spring clip ...*

▲ *Front sidelamp bulb is plugged into rear of headlamp*

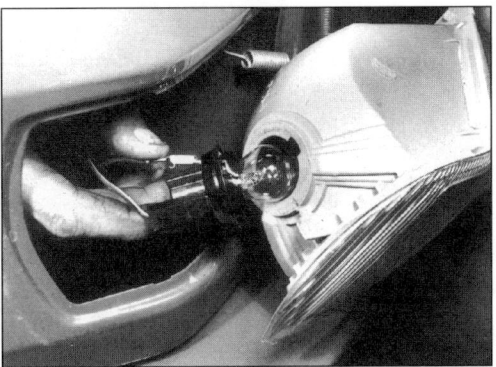

▲ *... so that lamp bulbholder can be removed when lamp is withdrawn from front of car*

Remove the bulbholder from the rear of the lamp by twisting it anti-clockwise.

Remove the bulb from the bulbholder by pressing it in and twisting it anti-clockwise.

Fit the new bulb using a reversal of the removal procedure; where the lamp was removed, ensure that its flange is located correctly behind the body panel, and that the lamp's seal is seated properly before fastening the spring clip.

FRONT DIRECTION INDICATOR SIDE REPEATER LAMP

Press the lamp to its right to release its left-hand retainer, then withdraw the lamp from the wing panel.

Rotate (anti-clockwise) and pull the bulbholder from the lamp body.

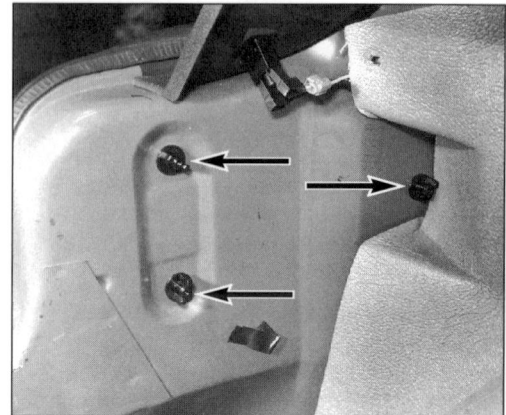

▲ #Unscrew three plastic nuts (arrowed) to release Maestro rear lamp cluster ...

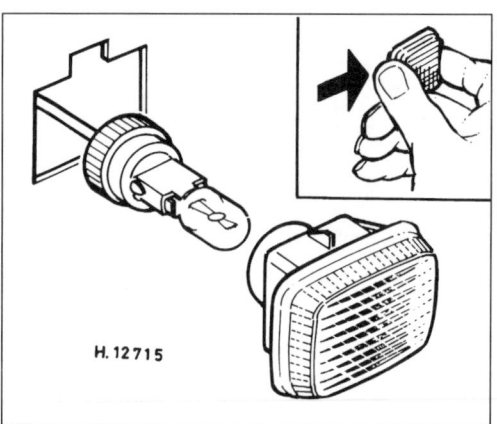

▲ Front direction indicator side repeater lamp bulb renewal

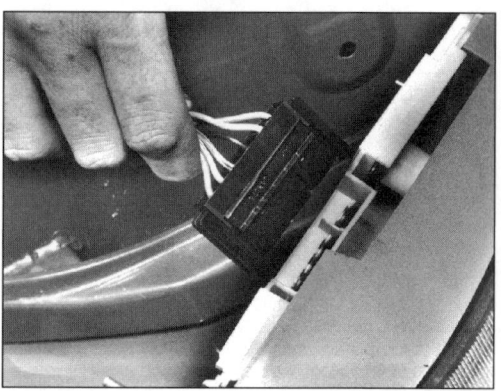

▲ ... withdraw cluster and unplug wiring connector ...

Pull the bulb from its socket; check carefully the condition of the bulb contacts (see introductory note at the start of this Section).

Push a new bulb into the socket, and refit the assembly in the reverse order to removal.

REAR LAMPS – IN CLUSTER

Maestro

Open the tailgate, release the rear edge of the luggage compartment side cover, then unscrew the three plastic nuts securing the lamp cluster to the body rear panel.

Withdraw the cluster, unplugging the wiring connector, then press the bulbholder panel

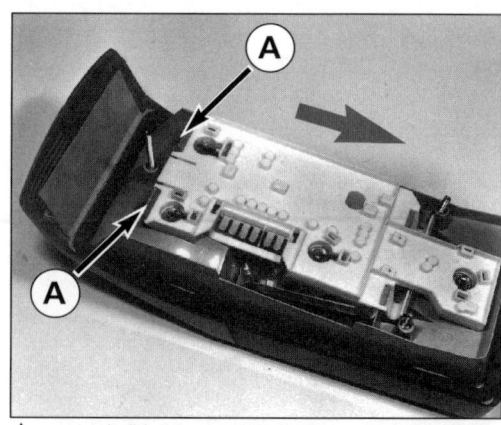

▲ ... press bulbholder panel in direction arrowed until catches 'A' are released ...

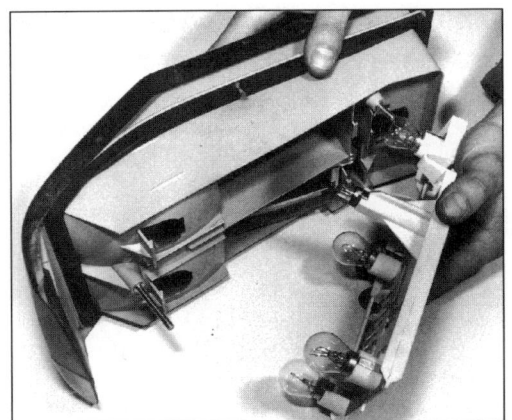

▲ .... and bulbholder panel can be removed from lamp lens

towards the two retaining studs until the two catches are released, and the panel can be separated from the lamp lens.

Remove the faulty bulb by pressing it in and twisting it anti-clockwise, except for the rear sidelamp bulb **(D),** which is pulled out of its holder. Check carefully the condition of the bulb contacts and the bulbholder panel printed circuit metal strips; if any corrosion is found on these, it must be cleaned off – in extreme cases, the panel must be renewed to cure

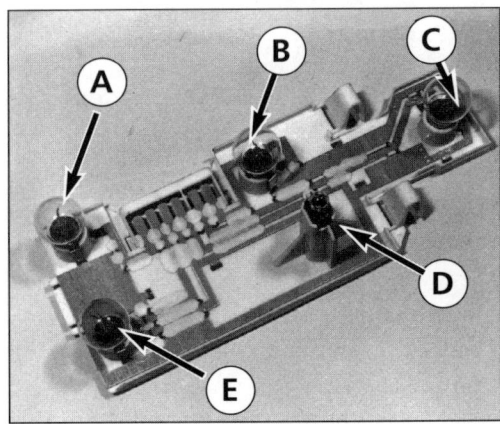

▲ Rear lamp bulb identification – Maestro
 A Direction indicator
 B Reversing lamp
 C Foglamp
 D Sidelamp
 E Brake/tail lamp

puzzling rear lamp faults (especially where the use of one lamp interferes with another's operation). Note that where one panel is found to be faulty, the other should be checked carefully and renewed at the same time, if necessary.

Fit the new bulb using a reversal of the removal procedure; the brake/tail lamp bulb **(E)** has offset pins, and so will only fit correctly one way.

Montego Saloon

Open the boot. Press and disengage the two retainers at the ends of the bulbholder panel, then withdraw the panel; unplug the wiring connector only if the extra space is required.

▲ Press retainer to disengage ...

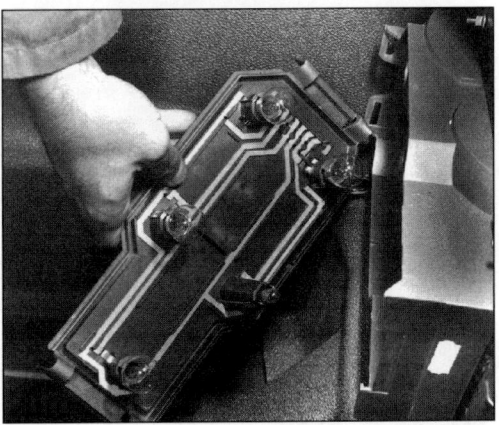

▲ ... so that bulbholder panel can be removed from rear lamp cluster lens – Montego Saloon

Bulb identification, removal and refitting details are all exactly the same as for Maestro models; the note concerning the checking of the bulbholder panel contacts applies equally.

Montego Estate

Open the tailgate and remove the cover from the luggage compartment stowage box on the appropriate side. Undo the retaining screws and withdraw the stowage box, then reach into the aperture and twist the relevant bulbholder anti-clockwise to release it.

Remove the bulb by pressing it in and twisting it anti-clockwise.

Fit the new bulb using a reversal of the removal procedure.

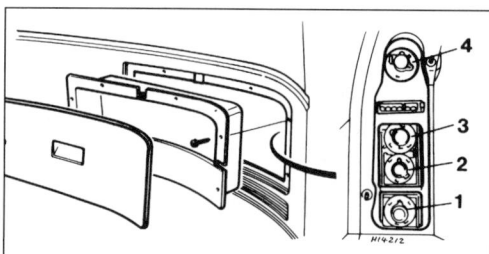

▲ *Rear lamp cluster access and bulb identification – Montego Estate*
  1 *Foglamp*
  2 *Reversing lamp*
  3 *Direction indicator*
  4 *Brake/tail lamp*

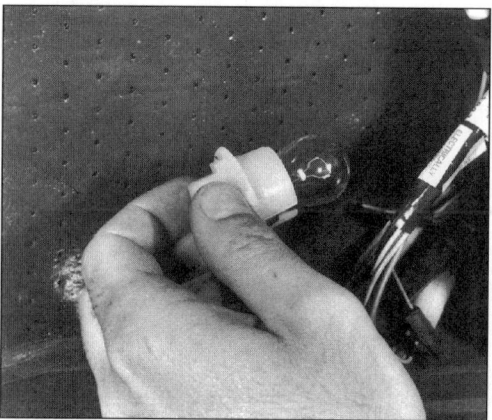

▲ *Rear lamp bulbholder removed – Montego Estate*

REAR NUMBER PLATE LAMP

Maestro

Using a small screwdriver, press in the catch on each side of the lamp lens and lift the lamp assembly out of the rear bumper.

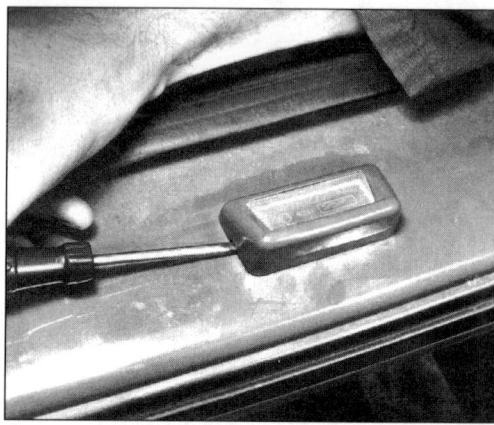

▲ *Depress catches to release rear number plate lamp from bumper – Maestro ...*

▲ *... depress retainers to separate lens from lamp body – early models*

On early models, depress the retainers to separate the lamp lens from the lamp body; remove the bulb by pressing it in and twisting it anti-clockwise.

On later models, rotate (anti-clockwise) and pull the bulbholder from the lamp body; pull the bulb from its socket.

▲ *Rear number plate lamp bulb renewal –
Montego saloon*

▲ *Undo lamp retaining screws (arrowed) ...*

▲ *... to reach rear number plate lamp bulb –
Montego Estate*

On all models, check carefully the condition of the bulb contacts (see introductory note at the start of this Section on page 127). Refit the assembly in the reverse order to removal.

Montego Saloon

Press the lamp assembly forwards to release it, then lift it out of the rear bumper, and twist anti-clockwise to separate the lens from the bulbholder.

The bulb is a capless/wedge type, and is removed by pulling it out of the holder.

Fit the new bulb using a reversal of the removal procedure, but be particularly careful to check the condition of the bulb contacts (see introductory note at the start of this Section on page 127).

Montego Estate

Unscrew the screw at each end of the lamp lens, then withdraw the lamp assembly and prise the bulb out of its spring contacts.

Fit the new bulb using a reversal of the removal procedure, but be particularly careful to check the condition of the bulb contacts (see introductory note at the start of this Section on page 127).

Interior light bulbs

Note: *Renewal of the various instrument panel and facia-mounted control illumination light bulbs requires detailed explanation, and in some cases extensive dismantling. Details can be found in the relevant Owners Workshop Manual for your particular car (OWM 922 for Maestro 1.3 & 1.6 models, OWM 1066 for Montego 1.3 & 1.6 models, or OWM 1067 for Montego 2.0 models).*

COURTESY AND LUGGAGE COMPARTMENT/BOOT LAMPS

Starting at the end opposite to the switch (where applicable), carefully prise the lamp assembly from its location, taking care not to damage the surrounding trim. Press the bulb in and twist to remove it.

Fit the new bulb, making sure that it locates securely in its contacts, then refit the lamp assembly.

▲ Courtesy and luggage compartment/boot lamp bulb renewal

GLOVEBOX LAMP

Open the glovebox lid, and use a small screwdriver to carefully prise the lamp assembly from its location, then pull the bulb from the lamp assembly's spring contacts.

Fit the new bulb, making sure that it locates securely in its contacts, then refit the lamp assembly.

FOOTWELL LAMP

Squeeze together the retainers and ease the lamp from its location. Pull the bulb from the lamp assembly's spring contacts.

Fit the new bulb, making sure that it locates securely in its contacts, then refit the lamp assembly.

CIGAR LIGHTER ILLUMINATION BULB

Front

Taking care not to mark the surrounding trim, carefully prise the sub-panel from the facia. Pull the bulbholder from the lighter housing, pull the bulb out of its socket and press in the new bulb.

Ensure that the bulbholder earth strip is against the lighter housing on refitting.

Rear

Move either front seat fully forwards, and carefully prise the trim panel from between the seat belt stalks, taking care not to damage the centre console. Pull the bulbholder from the lighter housing, pull the bulb out of its socket

and press in the new bulb.

Ensure that the bulbholder earth strip is against the lighter housing on refitting.

AUTOMATIC TRANSMISSION SELECTOR ILLUMINATION BULB

Maestro

Prise out the blanking plate, coin tray or ashtray from the front of the console, and undo the two screws beneath which secure the front of the console. Lift the console until the appropriate bulbholder can be withdrawn.

Pull the bulb out of its socket and press in the new bulb.

Refitting is the reverse of removal.

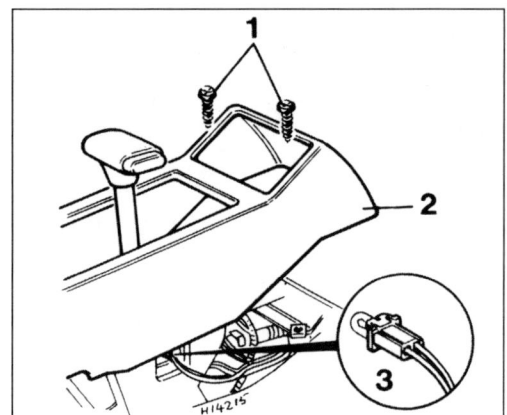

▲ Automatic transmission selector illuminating bulb renewal – Maestro
1 Console front securing screws
2 Centre console
3 Bulbholder

Montego

Moving the front seats fully forwards or rearwards as required, and taking care not to damage the centre console, carefully prise the trim panels from between the seat belt stalks and from in front of the handbrake lever; on early models, prise out the blanking plate, coin tray or ashtray from the front of the console. Undo the screws beneath these panels which secure the front and rear of the console, then slide it rearwards and lift it at the front until the appropriate bulbholder can be withdrawn.

Pull the bulb out of its socket and press in the new bulb.

Refitting is the reverse of removal.

HAZARD WARNING LAMP SWITCH

With the switch in the 'ON' position, pull off the switch lens and pull the bulb out of its socket.

Fit the new bulb using a reversal of the removal procedure.

▲ Fuse location – early Maestro models

▲ Hazard warning lamp switch bulb renewal

▲ Typical fuse (and relay) location – all other models

FUSES

Most of the car's electrical circuits are protected by fuses, which will blow when the relevant circuit becomes overloaded. This is to prevent possible damage to electrical components, or the possible risk of fire.

● **Before renewing a fuse, always make sure that the ignition and the relevant circuit are switched off.**

On early Maestro models, the fuses are housed in a panel in the top of the facia parcel shelf; to reach the fuses, remove the mat and lift the fusebox cover.

On all other models, the fuses are housed behind the bottom right-hand corner of the facia. Using a coin or similar, turn the two fasteners through a quarter-turn to release them and remove the fusebox cover to reach the fuses.

The fuse locations are numbered to identify them; the corresponding circuits are indicated on the fusebox cover label.

To check a fuse, remove the fusebox cover and use a pair of small pliers to pull out the appropriate fuse; on some models, a pair of fuse-removing tweezers are provided.

All fuses are a push fit in their sockets; a blown fuse can be recognised by a break in the metal strip between the two terminal blades. On fitting a fuse, ensure that its terminal blades enter between the fusebox panel terminals.

▲ *A blown fuse will be broken at the point arrowed*

If a fuse has blown, renew it with one of identical rating. The fuses are colour-coded to show their rating as follows:

Colour	Rating (amps)
Purple	3
Orange	5
Red	10
Blue	15
Clear	25
Green	30

Spare fuses are located as follows – on Maestros they are either in row 7 of the fusebox panel (early models), or in the fusebox cover (later models); on Montegos, they are either in a separate holder to one side of the fusebox panel (early models), or at one end of the numbered rows (later models). If a spare fuse is ever used, replace it as soon as possible to ensure that a spare of each rating is always carried in the car.

Always use proper fuses of the correct type and rating, and **never** use wire or any other material to bridge the gap where a fuse should be.

Note that the same fuse should never be renewed more than once without investigating the source of the trouble. Seek advice from someone suitably qualified if necessary.

Note that, depending on the unit and installation, the radio may have its own in-line circuit fuse (usually located in the wiring loom behind the radio), or may also be protected by a fuse in the rear of the radio casing.

RELAYS

Various types of relays are used in some circuits, and a faulty relay may prevent one or more components from working.

● **Before renewing a relay, always make sure that the ignition and the relevant circuit are switched off.**

Relays are of the plug-in type, most of the major relays being located in a panel located in the glovebox roof on early Maestro models, or around the fusebox panel (refer to 'Fuses' on page 135) on all other models. On most models, the circuit covered by each relay is indicated either on the panel itself, or on the fusebox cover label.

Various additional relays may be used depending on the equipment fitted, and these may be located in various positions depending on model. If a problem is suspected which may be due to a relay not located in the fusebox, it's best to seek advice from a Rover dealer, who will know where on the car to find the relevant components.

To remove a relay, simply pull it from its socket.

It's difficult to tell visually whether a relay is faulty, and the best test is to substitute a known good relay to see if the relevant circuit then operates – if it does, then it's probably the relay which was causing the problem.

THE MOT TEST

The following information is intended as a guide to enable you to spot some of the more obvious faults which may cause your car to fail the MOT test.

Obviously it isn't possible to check the car to the same standard as a professional MOT tester, who will be highly experienced and will have all the necessary tools and equipment.

Although we can't cover all the points of the test, the following should provide you with a good indication as to the general condition of the car, and will enable you to identify any obvious problem areas before submitting your car for the test.

Further explanation of most of the checks, and details of how to cure any problems discovered can be found in the relevant Owners Workshop Manual for your particular car (OWM 922 Maestro 1.3 & 1.6, OWM 1066 Montego 1.3 & 1.6, or OWM 1067 Montego 2.0).

Lights

● All lights must work, and the lenses and reflectors must not be damaged.
● Pairs of similar lights must be of the same brightness (eg both rear lights).
● Both headlights must show the same colour, and must be correctly aimed so that they light up the road adequately, but don't dazzle other drivers.
● The brake lights must work when the brake pedal is pressed.
● The direction indicator lights must flash between one and two times per second.

Steering and suspension

● All the components and their mounting points must be secure, and there must be no damage or excessive corrosion.
● There must be no excessive free play in the joints and bushes, and all rubber gaiters must be secure and undamaged.
● The steering mechanism must operate smoothly without excessive free play or roughness.
● There should be no excessive free play or roughness in the wheel bearings, but there should be sufficient free play to prevent tightness or binding.
● There should be no major fluid leaks from the shock absorbers and, when each corner of the car is depressed, it should rise and then settle in its normal position.
● The driveshafts and propeller shaft (on rear wheel drive cars) must not be damaged or distorted.

Brakes

● It should be possible to operate the handbrake without excessive force or excessive movement of the lever, and it must not be possible for the lever to release unintentionally. The handbrake must be able to lock the relevant wheels on which it operates.
● The brake pedal must not be damaged, and when the pedal is pressed, resistance should be felt near the top of its travel – hard resistance should be felt, and the pedal should not move down towards the floor. The resistance should be firm and not spongy.
● There should be no signs of any fluid leaks anywhere in the braking system.
● The wheels should turn freely when the brakes are not applied.
● When the brakes are applied, the brakes on all four wheels must work, and the car must stop evenly in a straight line without pulling to one side.
● All the braking system components must be secure, and there must be no signs of excessive wear or corrosion.

Tyres and wheels

● The tyres must be in good condition, and there must be no signs of excessive wear or damage (refer to *'Regular checks'* on page 85).
● Tyres at the same end of the car must be of the same size and type. Always fit radial tyres; crossply tyres are not suitable for modern cars.

Seatbelts

● The mountings must be secure, and must not be loose or excessively corroded.
● The seatbelt webbing should not be frayed or damaged (this must be checked along the full length of the belts).
● When a seatbelt is fastened, the locking mechanism should hold securely and should release when intended.
● In the case of inertia reel seatbelts, the retractors should work properly when the belts are released.

General

● The windscreen wipers and washers must work properly, and the wiper blades must clear the windscreen without smearing.
● The horn must work properly.
● The exhaust mountings must be secure, and the system itself must be free from leaks and serious corrosion.
● There must be no serious corrosion or damage to the vehicle structure. Corrosion of body panels will not necessarily cause a car to fail the test, but there should be no serious corrosion of any of the 'load bearing' structural components.
● The exhaust gas emissions level must be within certain limits depending on the age of the car (special equipment is required to measure exhaust gas emissions), and the MOT test also includes a visual check for excessive exhaust smoke when the engine is running. Generally, the car should meet the emission regulations if regular servicing has been carried out.

AUSTIN/MG MAESTRO & MONTEGO

The following table isn't intended as an exhaustive guide to fault finding, but it summarises some of the more common faults which may crop up during a car's life.

Hopefully, the table should help you to find the cause of a problem, even if you can't cure it yourself.

When confronted with a fault, try to think calmly and logically about the symptom(s), and you should be able to work out what the fault *can't* be! Check one item at a time, otherwise if you do clear the fault, you may not know what was causing it.

The commonest cause of difficulty is starting, especially in the Winter. Make sure that your battery is kept fully charged, and that the ignition components are in good condition, clean and dry (refer to *'Servicing'* on page 111).

Further details of fault diagnosis, along with comprehensive renewal procedures for most components, can be found in our Owners Workshop Manual for your particular model (OWM 922 for Maestro 1.3 & 1.6 models, OWM 1066 for Montego 1.3 & 1.6 models, or OWM 1067 for Montego 2.0 models).

SYMPTOM	POSSIBLE CAUSES
ENGINE	
Starter motor doesn't turn, and headlamps don't come on	● Flat battery ● Loose, dirty or corroded battery connections ● Other electrical or wiring fault
Starter motor doesn't turn, and headlamps dim	● Battery charge low ● Loose, dirty or corroded battery connections ● Faulty starter motor ● Other electrical or wiring fault ● Seized engine
Starter motor doesn't turn, and headlamps are bright	● Loose or dirty starter motor connections ● Faulty starter motor or ignition switch ● Gear selector lever not in position **N** or **P** (automatic transmission only) ● Other electrical or wiring fault
Starter motor spins, but doesn't turn engine	● Faulty starter motor ● Engine fault
Engine turns slowly but won't start	● Battery charge low ● Electrical or wiring fault ● Wrong grade of engine oil (refer to *'Service specifications'* on page 77)

SYMPTOM	POSSIBLE CAUSES

ENGINE (continued)

SYMPTOM	POSSIBLE CAUSES
Engine turns but won't fire, or engine starts but won't keep running	● Fuel tank empty ● Damp or dirty ignition system components ● Dirty or loose ignition system connections ● Incorrectly-adjusted or faulty spark plugs (refer to 'Servicing' on page 111) ● Fuel system fault ● Other electrical or wiring fault
Engine idles, but stalls when accelerator pedal is depressed	● Blocked or dirty air cleaner filter element (refer to 'Servicing' on page 109) ● Fuel system fault ● Ignition system fault ● Electrical fault
Poor acceleration, misfiring or lack of power	● Incorrectly-adjusted or faulty spark plugs (refer to 'Servicing' on page 111) ● Incorrect engine valve clearances ● Fuel system fault ● Ignition system fault ● Electrical fault
Engine continues to run when ignition is switched off	● Engine overheating (possibly due to low coolant level – refer to 'Regular checks' on page 84) ● Incorrectly-adjusted spark plugs or wrong type of spark plugs fitted (refer to 'Servicing' on page 111) ● Wrong grade of petrol ● Fuel system fault ● Excessive build-up of carbon inside engine ('decoke' required) ● Faulty ignition switch ● Other electrical or wiring fault
Engine doesn't reach normal operating temperature	● Faulty cooling system thermostat ● Faulty temperature gauge or sensor

SYMPTOM

POSSIBLE CAUSES

ENGINE (continued)

Engine overheats, or temperature gauge reads too high

- Airflow to radiator obstructed
- Blocked radiator, hose or engine coolant passage
- Coolant or engine oil level low (refer to *'Regular checks'* on page 83 or 84
- Coolant hose(s) leaking, worn or damaged
- Faulty cooling system thermostat
- Faulty water pump
- Faulty water pump drive
- Faulty temperature gauge or sensor
- Faulty electric cooling fan or switch

Ignition warning lamp lights when engine is running

- Alternator drivebelt loose or broken (refer to *'Servicing'* on page 106)
- Faulty alternator
- Other electrical or wiring fault

Oil pressure warning lamp lights when engine is running

- Oil level below 'MIN' mark on dipstick – refer to *'Regular checks'* on page 83 (if oil level is correct, and lamp is still on, seek advice, but **don't** start the engine)
- Oil leak
- Faulty oil pressure switch
- Badly-worn engine components

GEARBOX, TRANSMISSION AND CLUTCH

Difficulty in engaging gear (manual gearbox)

- Worn or faulty gearbox components
- Worn or faulty clutch mechanism
- Engine idle speed too high

Clutch slips (car does not accelerate when engine revs increase – manual gearbox)

- Faulty clutch mechanism
- Contaminated or worn clutch

Car gearchange is harsh (automatic transmission)

- Fault in transmission

Car pull away/gear change sluggish (automatic transmission)

- Low transmission fluid level (refer to *'Servicing'* on page 113)
- Fault in transmission

SYMPTOM	POSSIBLE CAUSES
BRAKES	
Brakes feel 'spongy'	● Air in brake hydraulic system
	● Fluid leak in brake system
Excessive brake pedal travel	● Faulty rear brake mechanism
	● Air in brake hydraulic system
	● Fluid leak in brake system
Brakes require excessive pedal pressure	● Damp, dirty or contaminated brake components
	● Fluid leak in brake system
	● Faulty or seized brake components
	● Faulty brake servo
Car pulls to one side	● Incorrect tyre pressure(s) or uneven tyre wear (refer to *'Regular checks'* on page 85)
	● Faulty or seized brake components
	● Worn or contaminated brake friction material
	● Incorrect wheel alignment
	● Worn or faulty steering components
	● Worn or faulty suspension components
Brakes squeal and/or judder	● Badly-worn or corroded brake components
	● Incorrectly-assembled brake components
	● Contaminated brake friction material
	● Worn or faulty suspension components
	● Worn or faulty steering components
Brake fluid level warning lamp lights	● Low brake fluid level (refer to *'Regular checks'* on page 84)
	● Faulty fluid level sensor or wiring

SYMPTOM | POSSIBLE CAUSES

SUSPENSION AND STEERING

Car becomes heavy to steer
- Low tyre pressure(s) (refer to *'Regular checks'* on page 85)
- Incorrect wheel alignment
- Worn or faulty steering components
- Worn or faulty suspension components

Car pulls to one side
- Refer to *'Brakes'* section of table

Car wanders
- Incorrect tyre pressure(s) or uneven tyre wear (refer to *'Regular checks'* on page 85)
- Car unevenly loaded
- Incorrect wheel alignment
- Worn or faulty suspension components
- Worn or faulty steering components

Car vibrates when driving
- Loose wheel bolt(s)
- Wheel(s) out of balance
- Worn or damaged driveshaft
- Worn or faulty suspension components
- Worn or faulty steering components
- Worn or faulty brake components

Hard or choppy ride
- Incorrect tyre pressure(s) (refer to *'Regular checks'* on page 85)
- Worn or faulty suspension components

Car leans excessively when cornering
- Roof rack overloaded
- Car unevenly loaded
- Worn or faulty suspension components

Uneven tyre wear
- Incorrect tyre pressure(s) (refer to *'Regular checks'* on page 85)
- Incorrect wheel alignment
- Wheel(s) out of balance
- Worn or faulty suspension components
- Worn or faulty steering components
- Brakes 'grabbing'

SYMPTOM	POSSIBLE CAUSES
ELECTRICS	
Electrical systems don't work	● Loose, dirty or corroded battery connections
	● Faulty earth connection
	● Battery charge low
	● Blown fuse (refer to *'Bulb, fuse and relay renewal'* on page 135)
	● Faulty relay
	● Fuse link, connecting main wiring loom to battery blown (seek advice)
Direction indicators don't work	● Blown fuse (refer to *'Bulb, fuse and relay renewal'* on page 135)
	● Faulty earth connection
	● Faulty direction indicator relay
	● Faulty direction indicator switch
Bulbs burn out repeatedly	● Faulty connections at lamp socket
	● Faulty alternator regulator
All lamps dim when engine speed drops to idle	● Loose alternator drivebelt (refer to *'Servicing'* on page 106)
	● Battery charge low
	● Faulty alternator

This Section should help you to understand some of the odd words of phrases used at garages and by 'car enthusiasts' which you may not be familiar with.

A

ABS – Abbreviation for Anti-lock Braking System. Uses sensors at each wheel to sense when the wheels are about to lock, and releases the brakes to prevent locking. This process occurs many times per second, and allows the driver to maintain steering control when braking hard.

Accelerator pump – A device attached to many *carburettors* which provides a spurt of extra fuel to the carburettor fuel/air mixture when the accelerator pedal is suddenly pressed down.

Additives – Compounds which are added to petrol and oil to improve their quality and performance.

Advance and retard – A system for altering the *ignition timing*.

AF – An abbreviation of 'Across Flats', the way in which many nuts, bolts and spanners are identified. AF is usually preceded by an Imperial unit of measurement – eg ½ in AF. Unless otherwise stated, all metric measurements are assumed to be AF, so the abbreviation is not normally used for metric nuts, bolts and spanners.

Air cooling – Alternative method of engine cooling in which no water is used. An engine-driven fan forces air at high speed over the surfaces of the engine.

ALB – Abbreviation for Anti-Lock Braking System. See *'ABS'*.

Alternator – A device for converting rotating mechanical energy into electrical energy. In modern cars, it has superseded the dynamo for charging the battery because of its much greater efficiency.

Ammeter – A device for measuring electrical current – the current supplied to the battery by the alternator, or drawn from the battery by the car's electrical systems.

Antifreeze – A chemical mixed with the water in the cooling system to lower the temperature at which the coolant freezes, and in modern cars to prevent corrosion of the metal in the cooling system.

Anti-roll bar – A metal bar mounted transversely across the car, connecting the two sides of the suspension, which counteracts the natural tendency for the car to lean when cornering.

Aquaplaning – A word used to describe the action of a tyre skating across the surface of water.

Automatic transmission – A type of *gearbox* which selects the correct gear ratio automatically according to engine speed and load.

Axle – Spindle on which a wheel revolves.

B

Balljoint – A ball-and-socket type joint used in steering and *suspension* systems, which allows relative movement in more than one plane.

Battery condition indicator – A device for measuring electrical voltage (a voltmeter) connected via the ignition switch to the car battery. Unlike an *ammeter*, it gives an indication that a battery is close to failing. **Also**, most 'maintenance-free' batteries have a battery condition indicator fitted to their casing, which consists of a small disc which changes its colour when the battery is close to failure and requires renewal.

Bearing – Metal or other hard wearing surface against which another part moves, and which is designed to reduce friction and wear (bearings are usually lubricated).

Bendix drive – A device on some types of starter motor which allows the motor to drive the engine for starting, then disengages when the engine starts to run.

BHP – see *Horsepower*.

Big end – The end of a *connecting rod* which is attached to the *crankshaft*. It incorporates a *bearing* and transmits the linear movement of the connecting rod to the crankshaft.

Bleed nipple (or valve) – A hollow screw which allows air or fluid to be bled out of a system when it is loosened.

Brake caliper – The part of a *disc brake* system which houses the *brake pads* and the hydraulically-operated pistons.

Brake disc – A rotating disc, coupled to a roadwheel, which is clamped between hydraulically operated friction pads in a *disc brake* system.

Brake fade – A temporary loss of braking efficiency due to overheating of the brake friction material.

Brake pad – The part of a *disc brake* system which consists of the friction material and a metal backing plate.

Brake shoe – The part of a *drum brake* system which consists of the friction material and a curved metal former.

Breather – A device which allows fresh air into a system or allows contaminated air out.

Bucket tappet – A bucket shaped component used in some engines to transfer the rotary movement of the *camshaft* to the up-and-down movement required for *valve* operation.

Bump stop – A hard piece of rubber used in many *suspension* systems to prevent the moving parts from contacting the body during violent suspension movements.

C

Camber angle – The angle at which the front wheels are set from the vertical, when viewed from the front of the car. Positive camber is the amount in degrees which the wheels are tilted out at the top.

Cam follower – A piece of metal used to transfer the rotary movement of the *camshaft* to the up-and-down movement required for *valve* operation.

Camshaft – A rotating shaft driven from the *crankshaft* with lobes or cams used to operate the engine *valves* via the *valve gear*.

Carbon leads – Ignition HT leads incorporating carbon (black fibres) which eliminates the need for separate radio and TV *suppressors*.

Carburettor – A device which is used to mix air and fuel in the proportions required for burning by the engine under all conditions of engine running.

Castor angle – The angle between the front wheels pivot points and a vertical line when viewed from the side of the car. Positive castor is when the axis is inclined rearwards.

Catalytic converter – A device incorporated in the exhaust system which speeds up the natural decomposition of the exhaust gases, and reduces the amount of harmful gases released into the atmosphere. Cars fitted with catalytic converters must be operated on *unleaded petrol*, as *leaded petrol* will destroy the catalyst.

Centrifugal advance – System of ignition *advance and retard* incorporated in many *distributors* in which weights rotating on a shaft alter the ignition timing according to engine speed.

Choke – This has two common meanings. It is used to describe the device which shuts off some of the air in a *carburettor* during cold starting (in order to provide extra fuel), and it may be manually or automatically operated. It's also used as a general term to describe a carburettor throttle bore.

Clutch – A friction device which allows two rotating components to be coupled together smoothly, without the need for either rotating component to stop moving.

Coil spring – A spiral coil of spring steel used in many *suspension* systems.

Combustion chamber – Shaped area in the *cylinder head* into which the fuel/air mixture is compressed by the *piston* and where the spark from the *spark plug* ignites the mixture.

Compression ratio (CR) – A term used to describe the amount by which the fuel/air mixture is compressed as a *piston* moves from the bottom to the top of its travel, and expressed as a number. For example an 8.5:1 compression ratio means that the volume of fuel/air mixture above the piston when the piston is at the bottom of its stroke is 8.5 times that when the piston is at the top of its stroke.

Compression tester – A special type of pressure gauge which can be screwed into a *spark plug* hole, which measures the pressure in the cylinder when the engine is turning but not firing. This gives an indication of engine wear or possible leaks.

Condenser (capacitor) – A device in a *contact breaker point distributor* which stores electrical energy and prevents excessive sparking at the contact breaker points.

Connecting rod ('con-rod') – Metal rod in the engine connecting a *piston* to the *crankshaft*.

Constant velocity (CV) joint – A joint used in *driveshafts*, where the instantaneous speed of the input shaft is exactly the same as the instantaneous speed of the output shaft at any angle of rotation. This does not occur in ordinary *universal joints*.

Contact breaker points – A device in the *distributor* which consists of two electrical points (or contacts) and a cam which opens and closes them to operate the *HT* electrical circuit which provides the spark at the *spark plugs*.

Crankcase – The area of the *cylinder block* below the *pistons* which houses the *crankshaft*.

Crankshaft – A cranked shaft which is driven by the *pistons* and provides the engine output to the *transmission*.

Crossflow cylinder head – A *cylinder head* in which the inlet and exhaust *valves* and *manifolds* are on opposite sides.

Crossply tyre – A tyre whose construction is such that the weave of the fabric material layers is running diagonally in alternately opposite directions to a line around the circumference of the tyre.

Cubic capacity – The total volume within the *cylinders* of an engine which is swept by the movement of the *pistons*.

CVH – A term applied by the Ford Motor Company to their overhead camshaft engines which incorporate a hemispherical *combustion chamber*. CVH means Compound Valve angle, Hemispherical combustion chamber.

Cylinder – Close fitting metal tube in which a *piston* slides. In the case of an engine, the cylinders may be bored directly into the *cylinder block*, or on some engines, cylinder liners are used which rest in the cylinder block and can be replaced when worn with matching pistons to avoid the requirement for *reboring* the cylinder block.

Cylinder block – The main engine casting which contains the *cylinders*, *crankshaft* and *pistons*.

Cylinder head – The casting at the top of the engine which contains the *valves* and associated operating components.

D

Damper – See *shock absorber*.

Dashpot – An oil-filled *cylinder* and *piston* used as a damping device in SU and Zenith/Stromberg CD type carburettors.

Dead axle (beam axle) – The simplest form of axle, consisting of a horizontal member attached to the car underbody by springs. This arrangement is used for the rear axle on some front-wheel-drive cars.

Decarbonising ('decoking') – Removal of all carbon deposits from the *combustion chambers* and the tops of the *pistons* and *cylinders* in an engine.

De Dion axle – A rear axle consisting of a cranked tube attached to the wheel hubs, with a separately mounted *differential* gear and *driveshafts*. *Suspension* is normally through *coil springs* between the wheel hubs and car underbody.

Derv – Abbreviation for Diesel-Engined Road Vehicle. A term often used to refer to Diesel fuel.

Diaphragm – A flexible membrane used in some components such as fuel pumps. The diaphragm spring used on *clutches* is similar, but is made from spring steel.

Diesel engine – An engine which relies on the heat generated when compressing air to ignite the fuel, and which therefore doesn't need an *ignition system*. Diesel engines have much higher *compression ratios* than petrol engines, normally around 20:1.

Differential – A system of gears (generally known as a crownwheel and pinion) which allows the *torque* provided by the engine to be applied to both driving wheels. The differential divides the torque proportionally between the driving wheels to allow one wheel to turn faster than the other, for example during cornering.

DIN – This stands for Deutsche Industrie Norm (German Industry Standard), which provides international standards for measuring engine power, torque, etc.

Disc brake – A brake assembly where a rotating disc is clamped between hydraulically operated friction pads.

Distributor – A device used to distribute the *HT* current to the individual *spark plugs*. The distributor may also contain the *advance and retard* mechanism. On some older cars, the distributor also contains the *contact breaker points* assembly.

Distributor cap – Plastic cap which fits on top of the *distributor* and contains electrodes, in which the *rotor arm* rotates to distribute the *HT* spark voltage to the correct *spark plug*.

DOHC – Abbreviation for Double Overhead Camshaft (see *'Twin-cam'*).

Doughnut – A term used to describe the flexible rubber coupling used in some *driveshafts*.

Driveshaft – Term usually used to describe the shaft (normally incorporating *universal* or *constant velocity joints*), which transmits drive from a *differential* to one wheel. More commonly found in front-wheel-drive cars.

Drive train – A collective term used to describe the *clutch/gearbox/transmission* and the other components used to transmit drive to the wheels.

Drum brake – A brake assembly with friction linings on 'shoes' running inside a cylindrical drum attached to the wheel.

Dual circuit brakes – A *hydraulic* braking system consisting of two separate fluid circuits, so that if one circuit becomes inoperative, braking power is still available from the other circuit.

Dwell angle – A measurement which corresponds to the number of degrees of *distributor* shaft rotation during which the *contact breaker points* are closed during the ignition cycle of one *cylinder*. The angle is altered by adjusting the contact breaker points gap.

E

Earth strap – A flexible electrical connection between the battery and a car earth point, or between the engine/*gearbox* and the car body to provide a return current path flow to the battery.

EFI – Abbreviation for Electronic *Fuel Injection*.

Electrode – An electrical terminal, eg in a *spark plug* or *distributor cap*.

Electrolyte – A current-conducting solution inside the battery (consisting of water and sulphuric acid in the case of a car battery).

Electronic ignition – An *ignition system* incorporating electronic components in place of *contact breaker points*, which can produce a much higher spark voltage than a contact breaker system, and is less affected by worn components.

Emission control – The reduction or prevention of the release into the atmosphere of poisonous fumes and gases from the engine and fuel system of a car. Required to different degrees by the laws of some countries, and achieved by engine design and the use of special devices and systems.

Epicyclic gears (planetary gears) – A gear system used in many *automatic transmissions* where there is a central 'sun' gear around which smaller 'planet' gears rotate.

Exhaust gas analyser – An instrument used to measure the amount of pollutants (mainly carbon monoxide) in a car's exhaust gases.

Expansion tank – A container used in many cooling systems to collect the overflow from the car's cooling system as the coolant heats up and expands.

F

Filter – A device for removing foreign particles from air, fuel or oil.

Final drive – A collective term (often expressed as a gear ratio) for the crownwheel and pinion (see *Differential*).

Flat engine – Form of engine design in which the *cylinders* are opposed horizontally, usually with an equal number on each side of the central *crankshaft*.

Float chamber – The part of a *carburettor* which contains a float and *needle valve* for controlling the fuel level in the reservoir.

Flywheel – A heavy rotating metal disc attached to the *crankshaft* and used to smooth out the pulsing from the *pistons*.

Four stroke (cycle) – A term used to describe the four operating strokes of a *piston* in a conventional car engine. These are (1) Induction – drawing the air/fuel mixture into the engine as the piston moves down; (2) Compression – of the fuel/air mixture as the piston rises; (3) Power stroke – where the piston is forced down after the fuel/air mixture has been ignited by the *spark plug*, and (4) Exhaust stroke – where the piston rises and pushes the burnt gases out of the *cylinder*. During these operations, the inlet and exhaust *valves* are opened and closed at the correct moment to allow the fuel/air mixture in, the exhaust gases out, or to provide a gas-tight compression chamber.

Fuel injection – A method of injecting fuel into an engine. Used in *Diesel engines* and also on some petrol engines in place of a *carburettor*.

Fuel injector – Device used on *fuel injection* engines to inject fuel directly or indirectly into the *combustion chamber*. Some fuel injection systems use a single fuel injector, while some systems use one fuel injector for each *cylinder* of the engine.

G

Gasket – Compressible material used between two surfaces to provide a leakproof joint.

Gearbox – A group of gears and shafts installed in a housing, positioned between the *clutch* and the *differential*, and used to keep the engine within its safe operating speed range as the speed of the car changes.

H

Half-shaft – A *driveshaft* used to transmit the drive from the *differential* to one of the rear wheels.

Hardy-Spicer joint (Hooke's or Cardan joint) – See *Universal joint*.

Helical gears – Gears in which the teeth are cut at an angle across the circumference of the gear to give a smoother mesh between gears and quieter running.

Horsepower – A measurement of power. Brake Horsepower (BHP) is a measure of the power required to stop a moving body.

HT – Abbreviation of High Tension (meaning high voltage) used to describe the *spark plug* voltage in an *ignition system*.

Hub carrier – A component usually found at each front corner of a car which carries the wheel and brake assembly, and to which the *suspension* and steering components are attached.

Hydraulic – A term used to describe the operation of a system by means of fluid pressure.

Hypoid gear – A gear with curved teeth which transmits drive through a right-angle, where the centreline of the drive gear is offset from the centreline of the driven gear. The meshing action of hypoid gears allows a larger and therefore stronger drive gear, and the meshing noise is reduced in comparison with conventional gears.

I

Independent suspension – A *suspension* system where movement of one wheel has no effect on the movement of the other, eg independent front suspension.

Ignition coil – An electrical coil which forms part of the *ignition system* and which generates the *HT* voltage.

Ignition system – The electrical system which provides the spark to ignite the air/fuel mixture in the engine. Normally the system consists of the battery, *ignition coil*, *distributor*, ignition switch, *spark plugs* and wiring.

Ignition timing – The time in the *cylinder* firing cycle at which the ignition spark (provided by the *spark plug*) occurs. The spark timing is normally a few degrees of *crankshaft* rotation before the *piston* reaches the top of its stroke, and is expressed as a number of degrees before top-dead-centre (BTDC).

Inertia reel – Automatic type of seat belt mechanism which allows the wearer to move freely in normal use, but which locks on sensing either sudden deceleration or a sudden movement of the wearer.

In-line engine – An engine in which the *cylinders* are positioned in one row as opposed to being in a *flat* or *vee* configuration.

J

Jet – A calibrated nozzle or orifice in a *carburettor* through which fuel is drawn for mixing with air.

Jump leads – Heavy electric cables fitted with clips to enable a car's battery to be connected to another battery for emergency starting.

K

Kerb weight – The weight of a car, unladen but ready to be driven, ie with enough fuel, oil, etc, to travel an arbitrary distance.

Kickdown – A device used on *automatic transmissions* which allows a lower gear to be selected for improved acceleration by fully depressing the accelerator.

Kingpin – A device which allows the front wheel of a car to swivel about a near vertical axis.

Knocking – See 'Pinking'.

L

Laminated windscreen – A windscreen which has a thin plastic layer sandwiched between two layers of toughened glass. It will not shatter or craze when hit.

Lead-free petrol – Contains no lead. It has no lead added during manufacture, and the natural lead content is refined out. This type of petrol is not currently available for general use in the UK, and should not be confused with *unleaded* petrol.

Leaded petrol – Normal 4-star petrol. Has a low amount of lead added during manufacture, in addition to the natural lead found in crude oil.

Leading shoe – A *drum brake* shoe of which the leading end (the one moved by the operating *pistons*) is reached first by a given point on the drum during normal forward rotation. A simple drum brake will have one leading and one trailing (the opposite) shoe.

Leaf spring – A spring commonly used on cars with a *live axle*, consisting of several long curved steel plates clamped together.

Limited slip differential – A type of *differential* which prevents one wheel from standing still while the other wheel spins excessively. Often used on high-performance cars.

Live axle – An axle through which power is transmitted to the rear wheels.

Loom – A complete car wiring system or section of a wiring system consisting of all the wires of correct length, etc, to wire up the various circuits.

LT – Abbreviation of Low Tension (meaning low voltage), used to describe battery voltage in the *ignition system*.

M

MacPherson strut – An independent front *suspension* system where the swivelling, springing and shock absorbing action of the wheels is dealt with by a single assembly.

Manifold – A device used for ducting the air/fuel mixture to the engine (inlet manifold), or the exhaust gases from the engine (exhaust manifold).

Master cylinder – A *cylinder* containing a *piston* and *hydraulic* fluid, directly coupled to a foot pedal (eg brake or *clutch* master cylinder). Used for transmitting pressure to the brake or clutch operating mechanism.

Metallic paint – Paint finish incorporating minute particles of metal to give added lustre to the colour.

Multigrade – Lubricating oil whose *viscosity* covers that of several single grade oils, making it suitable for use over a wider range of operating conditions.

N

Needle bearing – Type of *bearing* in which needle or cone-shaped rollers are used around the circumference to reduce friction.

Needle valve – A component of the *carburettor* which restricts the flow of fuel or fuel/air mixture according to the position of the valve in an orifice or *jet*.

Negative earth – Electrical system (almost universally adopted) in which the negative terminal of the car battery is connected to the car body. The polarity of all the electrical equipment is determined by this.

O

Octane rating – A scale rating for grading petrol.

ohc (overhead cam) – Describes an engine in which the *camshaft* is situated above the *cylinder head*, and operates the *valve gear* directly.

ohv (overhead valve) – Describes an engine which has its *valves* in the *cylinder head*, but with the *valve gear* operated by *pushrods* from a *camshaft* situated lower in the engine.

Oil cooler – Small *radiator* fitted in the oil circuit and positioned in a cooling airflow to cool the oil. Used mainly on high-performance engines.

Overdrive – A device coupled to a car's *gearbox* which raises the output gear ratio above the normal 1:1 of top gear.

Oversteer – A tendency for a car to turn more tightly into a corner than intended.

P

PCV (Positive Crankcase Ventilation) – A system which allows fumes and vapours which build up in the *crankcase* to be drawn into the engine for burning.

Pinion – A gear with a small number of teeth which meshes with one having a larger number of teeth.

Pinking – A metallic noise from the engine often caused by the *ignition timing* being too far *advanced*. The noise is the result of pressure waves which cause the *cylinder* walls to vibrate when the ignited fuel/air mixture is compressed.

Piston – Cylindrical component which slides in a closely-fitting metal tube or *cylinder* and transmits pressure. The pistons in an engine, for example, compress the fuel/air mixture, transmit the power to the *crankshaft*, and push the burnt gases out through the exhaust *valves*.

Piston ring – Hardened metal ring which is a spring fit in a groove running round the *piston* to ensure a gas-tight seal between the piston and *cylinder* wall.

Positive earth – The opposite of *negative earth*.

Power steering – A steering system which uses *hydraulic* fluid pressure (provided by an engine-driven pump) to reduce the effort required to steer the car.

Pre-ignition – See *'Pinking'*.

Propeller shaft – The shaft which transmits the drive from the *gearbox* to the rear axle in a front-engined rear-wheel-drive car.

Pushrod – A rod which is moved up and down by the rotary motion of the *camshaft* and operates the *rocker arms* in an *ohv* engine.

Q

Quarter light – A triangular window mounted in front or behind the main front or rear windows, usually in the front door, or behind the rear door.

Quartz-halogen bulb – A bulb with a quartz envelope (instead of glass), filled with a halogen gas. Gives a brighter, more even spread of light than an ordinary bulb.

R

Rack and pinion – Simplest form of steering mechanism which uses a *pinion* gear to move a toothed rack.

Radial ply tyre – A tyre in which the fabric material plies are arranged laterally, at right angles to the circumference.

Radiator – Cooling device through which the engine coolant is passed, situated in an air flow and consisting of a system of fine tubes and fins for rapid heat dissipation.

Radius arms (rods) – Locating arms sometimes used with a *live axle* to positively locate it in the fore-and-aft direction.

Rebore – The process of enlarging the *cylinder* bores to a very accurately specified measurement in order to fit new *pistons* to overcome wear in the engine. Not normally necessary unless the engine has covered a very high mileage.

Recirculating ball steering – A derivative of *worm and nut* steering, where the steering shaft motion is transmitted to the steering linkage by balls running in the groove of a worm gear.

Rev counter – See *Tachometer*.

Rocker arm – A lever which rocks on a central pivot, with one end moved up and down by the *camshaft*, and the other end operating an inlet or exhaust *valve*.

Rotary engine – See '*Wankel engine*'.

Rotor arm – A rotating arm in the *distributor* which distributes the *HT* spark voltage to the correct *spark plug*.

Running on – A tendency for an engine to keep on running after the ignition has been switched off. Often caused by a badly maintained engine or the use of an incorrect grade of fuel.

S

SAE – Society of Automotive Engineers (of America). Lays down international standards for the classification of engine performance and many other specifications, but is most commonly used to classify oils.

Safety rim – A special wheel rim shape which prevents a deflated tyre from rolling off the wheel.

Sealed beam – A sealed headlamp unit where the filament is an integral part and cannot be renewed separately.

Semi-trailing arm – A common form of independent rear *suspension*.

Servo – A device for increasing the normal effort applied to a control.

Shock absorber – A device for damping out the up-and-down movement of the *suspension* when the car hits a bump in the road.

Spark plug – A device with two *electrodes* insulated from each other by a ceramic material, which screws into an engine *combustion chamber*. When the *HT* voltage is applied to the plug terminal, a spark jumps across the electrodes and ignites the fuel/air mixture.

Squab – Another name for a seat cushion.

Steel-braced tyre – Tyre in which extra plies containing steel cords are incorporated with the fabric plies to give added strength.

Steering arm (knuckle) – A short arm on the front *hub carrier* to which the steering linkage connects.

Steering gear – A general term used to describe the steering components, usually refers to a steering rack-and-pinion assembly.

Steering rack – See *Rack and pinion*.

Stroboscopic light – A light switched on and off by the engine *ignition system* which is used for checking the *ignition timing* when the engine is running.

Stroke – The total distance travelled by a single *piston* in its *cylinder*.

Stub axle – A short axle which carries one wheel.

Subframe – A small frame which is mounted on the car's body, and carries the *suspension* and/or the *drivetrain* assemblies.

Sump – The main reservoir for the engine oil.

Supercharger – A device which uses an engine-driven turbine (usually driven by a belt or gears from the *crankshaft*) to drive a compressor which forces air into the engine, providing increased fuel/air mixture flow, and therefore increased engine efficiency. Sometimes used on high-performance engines.

Suppressor – A device which is used to reduce or eliminate electrical interference caused by the *ignition system* or other electrical components.

Suspension – A general term used to describe the components which suspend the car body on its wheels.

Swing axle – A *suspension* arm which is pivoted near the front-to-rear centreline of the car, and which allows the wheel to swing vertically about that pivot point.

Synchromesh – A device in a *gearbox* which synchronises the speed of one gear shaft with another to produce smooth, noiseless engagement of the gears.

T

Tachometer – Also known as a rev counter, indicates engine speed in revolutions per minute (rpm).

Tappet – A term often used to refer to the component which transmits the rotary *camshaft* movement to the up-and-down movement required for *valve* operation.

Thermostat – A device which is sensitive to changes in engine coolant temperature, and opens up an additional path for coolant to flow through the *radiator* (to increase the cooling) when the engine has warmed up.

Tie-rod – A rod which connects the *steering arms* to the *steering gear*.

Timing belt – Fabric or rubber belt engaging on sprocket wheels and driving the *camshaft* from the *crankshaft*.

Timing chain – Metal flexible link chain engaging on sprocket wheels and driving the *camshaft* from the *crankshaft*.

Timing marks – Marks normally found on the *crankshaft* pulley or the *flywheel* and used for setting the ignition firing point with respect to a particular *piston*.

Toe-in/toe-out – The amount by which the front wheels point inwards or outwards from the straight-ahead position when steering straight ahead.

Top Dead Centre (TDC) – The point at which a *piston* is at the top of its *stroke*.

Torque – The turning force generated by a rotating component.

Torque converter – A coupling where the driving *torque* is transmitted through oil. At low speeds there is very little transfer of torque from the input to the output. As the speed of the input shaft increases, the direction of fluid flow within a system of vanes changes, and torque from the input impeller is transferred to the output turbine. The higher the input speed, the closer the output speed approaches it, until they are virtually the same.

Torsion bar – A metal bar which twists about its own axis, and is used in some *suspension* systems.

Toughened windscreen – A windscreen which when hit, will shatter in a particular way to produce blunt-edged fragments or will craze over but remain intact. A zone toughened windscreen has a zone in front of the driver which crazes into larger parts to reduce the loss of visibility which occurs when toughened windscreens break, but is otherwise similar.

Track rod – See *Tie-rod*.

Trailing arm – A form of independent *suspension* where the wheel is attached to a swinging arm, and is mounted to the rear of the arm pivot.

Transaxle – A combined *gearbox*/axle assembly from which two *driveshafts* transmit the drive to the wheels.

Transmission – A general term used to describe some or all of the *drivetrain* components excluding the engine, most commonly used to describe automatic gearboxes.

Turbocharger – A device which uses a turbine driven by the engine exhaust gases to drive a compressor which forces air into the engine, providing increased fuel/air mixture flow, and therefore increased engine efficiency. Commonly used on high-performance engines.

Twin-cam – Abbreviation for twin overhead *camshafts* (see *'ohc'*). Used on engines with a *crossflow cylinder head*, usually with one camshaft operating the inlet *valves* and the other operating the exhaust valves. Gives improved engine efficiency due to improved fuel/air mixture and exhaust gas flow in the *combustion chambers*.

Two stroke (cycle) – A common term used to describe the operation of an engine where each downward *piston* stroke is a power stroke. The fuel/air mixture is directed to the crankcase where it's compressed by the descending piston and pumped into the *combustion chamber*. As the piston rises, the mixture is compressed and ignited, which forces the piston down. The burnt gases flow from the exhaust port, but the piston is now compressing another fuel/air mixture charge in the crankcase which repeats the cycle.

U

Understeer – A tendency for a car to go straight on when turned into a corner.

Universal joint – A joint that can swivel in any direction whilst at the same time transmitting *torque*. This type of joint is commonly used in *propeller shafts* and some *driveshafts*, but is not suitable for some applications because the input and output shaft speeds are not the same at all positions of angular rotation. The type in common use is known as a Hardy-Spicer, Hooke's or Cardan joint.

Unleaded petrol – Has no lead added during manufacture, but still has the natural lead content of crude oil. Generally available in the

UK, most modern cars can use this type of petrol, but seek advice first, as engine adjustments may be required. Engine damage can occur if unleaded petrol is used incorrectly. Not to be confused with *lead-free* petrol which is not currently available in the UK.

Unsprung weight – The part of the car which is not supported by the springs.

V

Vacuum advance – System of ignition *advance and retard* used in some *distributors* where the vacuum in the engine inlet *manifold* is used to act on a *diaphragm* which alters the *ignition timing* as the vacuum changes due to the throttle position.

Valve – A device which opens or closes to allow or stop gas or fluid flow.

Valve gear – A general term used for the components which are acted on by the *camshaft* in order to operate the *valves*.

16-valve – Term used to describe a four *cylinder* engine with four *valves* per cylinder (usually two inlet valves and two exhaust valves). Gives improved engine efficiency due to improved fuel/air mixture and exhaust gas flow in the *combustion chambers*.

Vee engine – An engine design in which the *cylinders* are set in two banks forming a 'V' when viewed from one end. A V8 for example consists of two rows of four cylinders each.

Venturi – A streamlined restriction in the *carburettor* throttle bore which causes a low pressure to occur; this sucks fuel into the air stream to form a vapour suitable for combustion.

Viscosity – A term used to describe the resistance of a fluid to flow. When associated with lubricating oil, it's given an *SAE* number, 10 being a very light oil and 140 being a very heavy oil.

Voltage regulator – A device which regulates the *alternator* output to a predetermined level. On most alternators the voltage regulator is an integral part of the alternator, and regulates the charging current as well as the voltage.

W

Wankel engine – A rotary engine which has a triangular shaped rotor which performs the function of the *pistons* in a conventional engine, and rotates in a housing shaped approximately like a broad-waisted figure of eight. Very few cars use this type of engine.

Wheel balancing – Adding small weights to the rim of a wheel so that there are no out-of-balance forces when the wheel rotates.

Wishbone – An 'A'-shaped *suspension* component, pivoted at the base of the 'A' and carrying a wheel at the apex. Normally mounted close to the horizontal.

Worm and nut steering – A steering system where the lower end of the steering column has a coarse screw thread on which a nut runs. The nut is attached to a spindle which carries the drop arm which, in turn, moves the steering linkage.

A comprehensive network of local radio stations now exists throughout the UK.

Most of these radio stations provide regularly updated reports on traffic flow and road conditions, which can be of great help to drivers. Listening to traffic reports will help you to avoid the inevitable 'jams' which occur during every day driving.

Some car radio/cassette players are now equipped with a 'Radio Data System' (RDS) which will automatically tune into special traffic information signals, interrupting normal radio or tape listening when bulletins are broadcast. An RDS-equipped radio may prove to be a worthwhile investment if you travel by car regularly, particularly when driving on business trips.

The following table provides details of all the local radio frequencies throughout the UK, region-by-region. Where stations broadcast on FM/VHF and AM/MW, it's suggested that the FM/VHF frequency is used wherever possible, as this will usually give better reception.

AREA	FM/VHF	AM/MW
AVON		
Bath		
BBC Bristol	104.6	1548
GWR FM	103.0	–
Bristol		
BBC Bristol	94.9	1548
Brunel Classic Gold	–	1260
Galaxy Radio	97.2	–
GWR FM	96.3	–
BEDFORDSHIRE		
Bedford		
BBC Bedfordshire	95.5	1161
Chiltern Radio	96.9	–
SuperGold	–	792
Luton		
BBC Bedfordshire	103.8	630
Chiltern Radio	97.6	–
SuperGold	–	828
BERKSHIRE		
Reading		
210 Classic Gold Radio	–	1431
210 FM	97.0	–
BBC Berkshire	104.4	–
Wokingham		
BBC Berkshire	104.1	–
BIRMINGHAM		
Birmingham		
BBC WM	95.6	1458
BRMB FM	96.4	–
Buzz FM	102.4	–
Xtra AM	–	1152
BORDERS		
Eyemouth		
Radio Borders	103.4	–
Peebles		
Radio Borders	103.1	–
Selkirk		
Radio Borders	96.8	–
BUCKINGHAMSHIRE		
Milton Keynes		
BBC Bedfordshire	104.5	630
Horizon Radio	103.3	–

AREA	FM/VHF	AM/MW
CAMBRIDGESHIRE		
Cambridge		
BBC Cambridge	96.0	1026
CN FM	103.0	–
Peterborough		
BBC Cambridge	95.7	1447
Hereward Radio	102.7	1332
CHANNEL ISLANDS		
Guernsey		
BBC Guernsey	93.2	1116
Jersey		
BBC Jersey	88.8	1026
CHESHIRE		
Echo 96	96.4	–
Chester		
BBC Merseyside	95.8	1485
Congleton		
Signal Cheshire	104.9	–
Macclesfield		
BBC Stoke	94.6	1503
Warrington		
BBC Merseyside	95.8	1485
CLEVELAND		
Middlesbrough		
BBC Cleveland	95.0	1548
TFM	96.6	–
GNR	–	1170
CLWYD		
Prestatyn		
BBC Cymru/Wales	94.2	882
Rhyl		
BBC Cymru/Wales	94.2	882
Wrexham		
BBC Cymru/Wales	93.3	657
MFM 1034	97.1/103.4	–
Marcher Gold	–	1260
CORNWALL		
Isles of Scilly		
BBC Cornwall	96.0	–
Liskeard		
BBC Cornwall	95.2	657
Redruth		
BBC Cornwall	103.9	630
CUMBRIA		
Barrow-in-Furness		
BBC Furness	96.1	837
Kendal		
BBC Cumbria	95.2	–
Windermere		
BBC Cumbria	104.2	–
Workington		
BBC Cumbria	95.6	1458

AREA	FM/VHF	AM/MW
DERBYSHIRE		
Chesterfield		
BBC Sheffield	94.7	1035
Derby		
BBC Derby	94.2	1116
Trent FM	102.8	–
Gem AM	–	945
Matlock		
BBC Derby	95.3	–
DEVON		
Barnstaple		
BBC Devon	94.8	801
Exeter		
BBC Devon	95.8	990
DevonAir	97.0	666/954
South West	103.0	–
Okehampton		
BBC Devon	96.0	801
Plymouth Area		
BBC Devon	103.4	855
Plymouth Sound	97.0	1152
Tavistock		
Radio in Tavistock	96.6	–
Torbay		
BBC Devon	103.4	1458
DevonAir	96.4	666/954
DORSET		
Poole		
BBC Solent	96.1	1359
Bournemouth		
BBC Solent	96.1	1359
2CR Classic Gold		828
2CR FM	102.3	–
Weymouth		
BBC Solent	96.1	–
DUMFRIES & GALLOWAY		
Dumfries		
BBC Scotland/Solway	94.7	585
SW Sound	97.2	–
Stranraer		
BBC Scotland	94.7	
DURHAM		
Darlington		
BBC Newcastle	95.4	–
Durham		
BBC Newcastle	95.4	1458
DYFED		
Aberystwyth		
BBC Cymru/Wales	93.1	882

AREA	FM/VHF	AM/MW
EAST SUSSEX		
Brighton		
BBC Sussex	95.3	1485
South Coast Radio	–	1323
Southern Sound Classic Hits	103.5	–
Eastbourne		
BBC Sussex	104.5	1161
Southern Sound Classic Hits	102.4	–
Hastings		
Southern Sound Classic Hits	97.5	–
Newhaven		
Southern Sound Classic Hits	96.9	–
ESSEX		
Basildon		
BBC Essex	95.3	765
Chelmsford		
BBC Essex	103.5	765
Breeze AM		1431/1359
Essex Radio	102.6	–
Colchester		
BBC Essex	103.5	729
Mellow 1557	–	1557
Harlow		
BBC Essex	–	765
Manningtree		
BBC Suffolk	103.9	–
Southend-on-Sea		
BBC Essex	95.3	1530
Breeze AM		1431/1359
Essex Radio	96.3	–
FIFE		
Kirkcaldy		
BBC Scotland	94.3	810
GLOUCESTERSHIRE		
Gloucester		
BBC Gloucester	104.7	603
Severn Sound	102.4	
Three Counties Radio		774
Stroud		
BBC Gloucester	95.0	603
Severn Sound	103.0	–
GRAMPIAN		
Aberdeen		
BBC Scotland/Aberdeen	93.1	990
North Sound	96.9	1035
Stirling		
Central FM	96.7	–

AREA	FM/VHF	AM/MW
GWENT		
Newport		
Red Dragon Radio	97.4	–
Touch AM		1305
GWYNEDD		
Anglesey		
BBC Cymru/Wales	94.2	882
HAMPSHIRE		
Andover		
210 FM	102.9	–
Basingstoke		
210 FM	102.9	–
BBC Berkshire	104.1	–
Isle of Wight		
BBC Solent	96.1	999/1359
Isle of Wight Radio	–	1242
Portsmouth		
BBC Solent	96.1	999
Oceansound	97.5	–
South Coast Radio	–	1170
Southampton		
BBC Solent	96.1	999
Oceansound	97.5	–
Power FM	103.2	–
South Coast Radio	–	1557
Winchester		
Oceansound	96.7	–
Power FM	103.2	
HEREFORDSHIRE		
Hereford		
BBC Here/Worc	94.7	819
Radio Wyvern	97.6	954
HERTFORDSHIRE		
Watford		
BBC GLR	94.9	1458
St Albans		
BBC Bedfordshire	103.8	630
Stevenage		
BBC Bedfordshire	103.8	630
HIGHLAND		
Inverness		
BBC Scotland	94.0	810
Moray Firth Radio	97.4	1107
Fort William		
BBC Scotland	93.7	–
Oban		
BBC Scotland	93.3	–
Thurso		
BBC Scotland	94.5	–

AREA	FM/VHF	AM/MW
HUMBERSIDE		
Kingston upon Hull Area		
BBC Humberside	95.9	1485
Classic Gold	–	1161
Viking FM	96.9	–
ISLE OF MAN		
Douglas		
Manx Radio	97.2	1368
Snaefell		
Manx Radio	89.0	–
KENT		
Ashford		
Invicta FM	96.1	–
Canterbury		
BBC Kent	104.2	774
Invicta FM	102.8	–
Dover		
BBC Kent	104.2	774
Invicta FM	97.0	–
East Kent		
Coast Classics	–	603
Folkestone		
BBC Kent	104.2	774
Maidstone & Medway		
BBC Kent	96.7	1035
Coast Classics	–	1242
Invicta FM	103.1	–
Royal Tunbridge Wells		
BBC Kent	96.7	1602
Thanet		
Invicta FM	95.9	–
LANCASHIRE		
Burnley Area		
BBC Lancashire	95.5	855
Blackpool		
BBC Lancashire	103.9	855
Red Rose Gold	–	999
Red Rose Rock FM	97.4	–
Lancaster		
BBC Lancashire	104.5	1557
Preston		
BBC Lancashire	103.9	855
Red Rose Gold	–	999
Red Rose Rock FM	97.4	–
LEICESTERSHIRE		
Leicester		
BBC Leicester	104.9	837
Gem AM	–	1260
Sound FM	103.2	–

AREA	FM/VHF	AM/MW
LINCOLNSHIRE		
Lincoln Area		
BBC Lincoln	94.9	1368
Lincs FM	102.2	–
LONDON		
Brixton		
Choice FM	96.9	–
Ealing		
Sunrise Radio	–	1413
Greater London		
BBC GLR	94.9	1458
Capital FM	95.8	–
Capital Gold	–	1548
Jazz FM	102.2	–
Kiss FM	100.0	–
LBC Newstalk	97.3	–
Melody Radio	104.9	–
Spectrum Int. Radio	–	558/990
Haringey		
London Greek Radio	103.3	–
WNK	103.3	–
Thamesmead		
London Talkback Radio	–	1152
RTM	103.8	–
Southall		
Sunrise Radio	–	1413
LOTHIAN		
Bathgate		
Radio Forth	97.3	–
Dunfermline		
BBC Scotland	94.3	810
Edinburgh		
BBC Scotland	94.3	810
Max AM	–	1548
Radio Forth	97.3/97.6	–
MANCHESTER		
Manchester		
BBC GMR	95.1	1458
KFM	104.9	–
Piccadilly Gold	–	1152
Piccadilly Key 103	103.0	–
Sunset Radio	102.0	–
MERSEYSIDE		
Liverpool		
BBC Mersey	95.8	1485
City Talk	–	1548
Radio City	96.7	–
MID GLAMORGAN		
Aberdare		
BBC Cymru/Wales	93.6	882
Merthyr Tydfil		
BBC Cymru/Wales	96.8	882

AREA	FM/VHF	AM/MW
MIDDLESEX		
Hounslow		
Sunrise Radio	–	1413
NORFOLK		
King's Lynn		
BBC Norfolk	104.4	873
Norwich Area		
BBC Norfolk	95.1	855
Broadland	102.4	1152
NORTHAMPTONSHIRE		
Kettering		
KCBC	–	1530
Northampton		
BBC Northampton	104.2	1107
Northants	96.6	–
SuperGold	–	1557
NORTHUMBERLAND		
Berwick-upon-Tweed		
BBC Newcastle	96.0	–
Radio Borders	97.5	–
NOTTINGHAM		
Nottingham		
BBC Nottingham	103.8	1521
Trent FM	96.2/96.5	–
Gem AM	–	999
ORKNEY		
Kirkwall		
BBC Scotland	93.7	–
OXFORDSHIRE		
Banbury		
Fox FM	97.4	–
Oxford		
BBC Oxford	95.2	1485
Fox FM	102.6	–
POWYS		
Welshpool		
BBC Cymru/Wales	94.0	882
SHETLAND		
Lerwick		
BBC Scotland	92.7	–
SIBC	96.2	–
SHROPSHIRE		
Ludlow		
BBC Shropshire	95.0	1584
RFM	97.6	–
Shrewsbury		
BBC Shropshire	96.0	–
Beacon Radio	97.2/103.1	
WABC	–	1017
Telford		
BBC Shropshire	96.0	756
Beacon Radio	97.2/103.1	
WABC	–	1017

AREA	FM/VHF	AM/MW
SOMERSET		
Mendip Area		
Orchard FM	102.6	–
Wells		
BBC Bristol	95.5	1548
SOUTH GLAMORGAN		
Cardiff		
BBC Cymru/Wales	96.8	882
Red Dragon Radio	103.2	–
Touch AM		1359
STAFFORDSHIRE		
Stafford		
Echo 96	96.9	–
Stoke on Trent		
BBC Stoke	94.6	1503
Signal Radio	102.6	1170
STRATHCLYDE		
Ayr		
West Sound	96.7	1035
Girvan		
West Sound	97.5	
Glasgow		
BBC Scotland	94.3	810
Clyde 1	102.5	–
Clyde 2	–	1152
Radio Clyde FM	97.3	
Greenock		
BBC Scotland	94.3	810
Kilmarnock		
BBC Scotland	93.9	810
SUFFOLK		
Bury St. Edmunds		
Saxon Radio	96.4	1251
Great Barton		
BBC Suffolk	104.6	–
Ipswich		
BBC Suffolk	103.9	–
Radio Orwell	97.1	1170

AREA	FM/VHF	AM/MW
SURREY		
Guildford		
BBC Surrey	104.6	–
Delta Radio	97.1	–
First Gold Radio	–	1476
Premier Radio	96.4	–
Reigate		
Radio Mercury	102.7	1521
TAYSIDE		
Dundee		
Radio Tay	102.8	1161
BBC Scotland	92.7	810
Perth		
Radio Tay	96.4	1584
TYNE & WEAR		
Fenham		
BBC Newcastle	104.4	–
Metro FM	103.0	–
Newcastle Area		
BBC Newcastle	95.4	1458
GNR	–	1152
Metro FM	97.1	–
Sunderland		
Wear FM	103.4	–
ULSTER		
Belfast		
BBC Ulster	94.5	1341
Classic Trax BCR	96.7	–
Cool FM	97.4	–
Downtown Radio	102.6	–
Enniskillen		
BBC Ulster	93.8	873
Downtown Radio	96.6	–
Limavady		
Downtown Radio	96.4	–
Londonderry		
BBC Foyle	93.1	792
Downtown Radio	102.4	–
WARWICKSHIRE		
Leamington Spa		
Mercia FM	102.9	–

AREA	FM/VHF	AM/MW
WEST GLAMORGAN		
Swansea		
BBC Cymru/Wales	93.9	882
Swansea Sound	96.4	1170
WEST MIDLANDS		
Coventry		
BBC CWR	94.8	–
Mercia FM	97.0	–
Radio Harmony	102.6	–
Xtra AM	–	1359
Sutton Coldfield		
BBC Derby	104.5	–
Beacon Radio	97.2/103.1	–
Wolverhampton		
BBC WM	95.6	828
Beacon Radio	97.2/103.1	–
WABC	–	990
WEST SUSSEX		
Crawley		
Radio Mercury	102.7	1521
Horsham		
BBC Sussex	95.1	1368
Worthing		
BBC Sussex	95.3	1485
WESTERN ISLES		
Stornoway (Lewis)		
BBC Scotland	94.2	–
WILTSHIRE		
Chippenham		
BBC Wiltshire Sound	104.3	–
Marlborough		
GWR FM	96.5	–
Salisbury		
BBC Wiltshire Sound	103.5	–
Swindon		
BBC Wiltshire Sound	103.6	1368
Brunel Classic Gold	–	1161
GWR FM	97.2	–
West Wilts		
Brunel Classic Gold	–	936
GWR FM	102.2	–
WORCESTERSHIRE		
Worcester		
BBC Here/Worc	104.0	738
Radio Wyvern	102.8	1530

AREA	FM/VHF	AM/MW
YORKSHIRE NORTH		
Northallerton		
BBC York	104.3	–
Scarborough		
BBC York	95.5	1260
Whitby		
BBC Cleveland	95.8	–
York		
BBC York	103.7	666
YORKSHIRE SOUTH		
Barnsley		
Classic Gold	–	1305
Hallam	102.9	–
Doncaster		
BBC Sheffield	104.1	1035
Classic Gold	–	990
Hallam	103.4	–
Rotherham		
BBC Sheffield	88.6	1035
Hallam	96.1	–
Sheffield		
BBC Sheffield	88.6	1035
Classic Gold	–	1548
Hallam	97.4	–
YORKSHIRE WEST		
Bradford		
Classic Gold	–	1278
Pennine	97.5	–
Sunrise FM	103.2	–
Halifax		
Classic Gold	–	1530
Pennine	102.5	–
Huddersfield		
Classic Gold	–	1530
Pennine	102.5	–
Leeds		
Aire FM	96.3	–
BBC Leeds	92.4	774
Magic 828	–	828

CONVERSION FACTORS

Length (distance)

Inches *(in)*	x 25.4	= Millimetres *(mm)*	x 0.0394	= Inches *(in)*
Feet *(ft)*	x 0.305	= Metres *(m)*	x 3.281	= Feet *(ft)*
Miles	x 1.609	= Kilometres *(km)*	x 0.621	= Miles

Volume (capacity)

Cubic inches *(cu in; in³)*	x 16.387	= Cubic centimetres *(cc; cm³)*	x 0.061	= Cubic inches *(cu in; in³)*
Pints *(pt)*	x 0.568	= Litres *(l)*	x 1.76	= Pints *(pt)*
Quarts *(qt)*	x 1.137	= Litres *(l)*	x 0.88	= Quarts *(qt)*
Gallons *(gal)*	x 4.546	= Litres *(l)*	x 0.22	= Gallons *(gal)*

Mass (weight)

Ounces *(oz)*	x 28.35	= Grams *(g)*	x 0.035	= Ounces *(oz)*
Pounds *(lb)*	x 0.454	= Kilograms *(kg)*	x 2.205	= Pounds *(lb)*

Force

Pounds-force *(lbf; lb)*	x 4.448	= Newtons *(N)*	x 0.225	= Pounds-force *(lbf; lb)*
Newtons *(N)*	x 0.1	= Kilograms-force *(kgf; kg)*	x 9.81	= Newtons *(N)*

Pressure

Pounds-force per square inch *(psi; lbf/in²; lb/in²)*	x 0.070	= Kilograms-force per square centimetre *(kgf/cm²; kg/cm²)*	x 14.223	= Pounds-force per square inch *(psi; lbf/in²; lb/in²)*
Pounds-force per square inch *(psi; lbf/in²; lb/in²)*	x 0.068	= Atmospheres *(atm)*	x 14.696	= Pounds-force per square inch *(psi; lbf/in²; lb/in²)*
Pounds-force per square inch *(psi; lbf/in²; lb/in²)*	x 0.069	= Bars	x 14.5	= Pounds-force per square inch *(psi; lbf/in²; lb/in²)*

Torque (moment of force)

Pounds-force feet *(lbf ft; lb ft)*	x 0.138	= Kilograms-force metres *(kgf m; kg m)*	x 7.233	= Pounds-force feet *(lbf ft; lb ft)*
Pounds-force feet *(lbf ft; lb ft)*	x 1.356	= Newton metres *(Nm)*	x 0.738	= Pounds-force feet *(lbf ft; lb ft)*
Newton metres *(Nm)*	x 0.102	= Kilograms-force metres *(kgf m; kg m)*	x 9.804	= Newton metres *(Nm)*

Power

Horsepower *(hp)*	x 0.745	= Kilowatts *(kW)*	x 1.3	= Horsepower *(hp)*

Velocity (speed)

Miles per hour *(miles/hr; mph)*	x 1.609	= Kilometres per hour *(km/h; kph)*	x 0.621	= Miles per hour *(miles/hr; mph)*

Fuel consumption*

Miles per gallon *(mpg)*	x 0.354	= Kilometres per litre *(km/l)*	x 2.825	= Miles per gallon *(mpg)*

Temperature

Degrees Fahrenheit = (°C x 1.8) + 32

Degrees Celsius (Degrees Centigrade; °C) = (°F − 32) x 0.56

* Note: It is common practice to convert from miles per gallon (mpg) to litres/100 kilometres (l/100 km), where mpg x l/100 km = 282

CONVERSION FACTORS

	LONDON	Aberdeen	Aberystwyth	Birmingham	Bournemouth	Brighton	Bristol	Cambridge	Cardiff	Carlisle	Chester	Derby	Dover	Edinburgh	Exeter	Fishguard	Fort William	Glasgow	Harwich	Holyhead
LONDON		806	341	179	169	89	183	84	249	484	298	198	117	608	275	422	818	634	114	420
Aberdeen	501		724	658	903	895	789	742	792	349	571	634	924	198	911	795	254	241	848	692
Aberystwyth	212	450		193	328	425	206	354	177	375	153	222	459	526	325	90	708	525	460	182
Birmingham	111	409	120		237	267	132	161	163	309	117	63	296	460	259	283	642	459	267	246
Bournemouth	105	561	204	147		148	124	251	195	554	351	306	278	705	132	375	887	703	283	463
Brighton	55	556	264	166	92		217	180	288	573	386	286	130	697	275	468	906	723	203	509
Bristol	114	490	128	82	77	135		233	71	439	237	195	314	591	121	251	722	589	314	340
Cambridge	52	461	220	100	156	112	145		290	417	269	155	204	544	351	436	750	566	106	391
Cardiff	155	492	110	101	121	179	44	180		443	237	225	385	587	192	180	776	608	359	341
Carlisle	301	217	233	192	344	356	273	259	275		222	304	602	151	571	444	333	150	523	341
Chester	185	355	95	73	218	240	147	167	147	138		114	415	373	357	243	555	372	375	137
Derby	123	394	138	39	190	178	121	96	140	189	71		315	436	315	312	637	454	261	251
Dover	73	574	285	184	173	81	195	127	239	374	258	196		726	391	539	935	752	203	538
Edinburgh	378	123	327	286	438	433	367	338	365	94	232	271	451		723	616	233	72	650	489
Exeter	171	566	202	161	82	171	75	218	119	355	222	196	243	449		372	904	721	389	460
Fishguard	262	494	56	176	233	291	156	271	112	276	151	194	335	383	231		777	594	533	272
Fort William	508	158	440	399	551	563	480	466	482	207	345	396	581	145	562	483		183	856	671
Glasgow	394	150	326	285	437	449	366	352	378	93	231	282	467	45	448	369	114		673	488
Harwich	71	527	286	166	176	126	195	66	223	325	233	162	126	404	242	331	532	418		497
Holyhead	261	430	113	153	288	316	211	243	212	212	85	156	334	304	286	169	417	303	309	
Hull	168	348	229	143	255	223	226	124	244	155	132	98	251	225	301	285	362	250	181	217
Inverness	537	105	482	445	594	590	528	497	524	253	387	430	601	159	603	529	66	176	563	463
Leeds	190	322	173	111	261	251	206	148	212	115	78	74	269	199	268	229	322	210	214	163
Leicester	98	417	151	39	166	153	112	68	139	208	94	28	171	294	187	207	415	301	134	183
Lincoln	136	382	186	85	217	191	163	86	190	180	123	51	213	259	238	245	387	273	152	208
Liverpool	205	334	104	94	235	260	164	184	164	116	17	88	278	210	239	160	323	211	250	94
Manchester	192	332	133	81	228	247	167	154	172	115	38	58	265	209	242	189	322	210	220	123
Newcastle	271	230	270	209	348	326	295	231	310	58	175	164	344	107	370	326	257	143	297	260
Norwich	107	487	267	155	212	162	221	60	237	287	210	139	169	364	277	331	494	380	63	299
Nottingham	123	388	154	50	189	178	132	84	152	185	87	16	196	265	207	210	392	278	150	172
Oxford	56	464	157	62	92	99	66	79	107	256	128	98	129	341	139	207	461	340	134	202
Penzance	282	683	313	268	193	269	186	329	230	466	333	307	354	560	111	342	673	559	353	397
Plymouth	213	614	244	199	124	213	117	260	161	397	264	238	285	491	42	273	604	490	284	328
Preston	216	302	144	105	256	271	185	198	190	85	49	100	289	179	266	200	292	180	264	125
Sheffield	162	358	158	75	227	217	163	124	176	145	78	36	235	235	248	214	352	240	190	163
Southampton	79	530	202	128	32	60	75	136	119	322	194	164	149	416	111	231	520	415	150	283
Stranraer	410	235	342	301	453	465	382	368	384	109	247	298	483	133	464	385	199	85	434	321
Swansea	196	506	76	126	162	220	85	216	41	289	151	165	280	383	160	71	496	382	264	189
Worcester	114	428	98	26	131	168	60	116	75	211	88	65	187	305	135	155	418	304	182	152
York	196	309	197	134	273	251	215	157	242	116	102	89	269	186	290	253	323	211	223	187

MILES

KILOMETRES

Hull	Inverness	Leeds	Leicester	Lincoln	Liverpool	Manchester	Newcastle	Norwich	Nottingham	Oxford	Penzance	Plymouth	Preston	Sheffield	Southampton	Stranraer	Swansea	Worcester	York	
270	864	306	158	219	330	309	436	172	198	90	454	343	348	261	127	660	315	183	315	**LONDON**
560	169	518	671	615	538	534	370	784	624	747	1099	988	486	576	853	378	814	689	497	Aberdeen
369	776	278	243	304	167	214	435	430	248	253	504	393	232	254	325	550	122	158	317	Aberystwyth
230	716	179	63	137	151	130	336	249	80	100	431	320	169	121	206	484	203	42	216	Birmingham
410	956	420	267	349	378	367	560	341	304	148	311	200	412	365	51	729	261	211	439	Bournemouth
359	950	404	246	307	418	398	525	261	286	159	433	343	436	349	97	748	354	270	404	Brighton
364	850	332	180	262	264	269	475	356	212	106	299	188	298	262	121	615	137	97	346	Bristol
200	800	238	109	138	296	248	372	97	135	127	529	418	319	200	219	592	348	187	253	Cambridge
393	843	341	224	306	264	277	499	381	245	172	370	259	306	283	192	618	66	121	389	Cardiff
249	407	185	335	290	187	185	93	462	298	412	750	639	137	233	518	175	465	340	187	Carlisle
212	623	126	151	198	27	61	282	338	140	206	536	425	79	126	312	398	243	142	164	Chester
158	692	119	45	82	142	93	264	224	26	158	494	383	161	58	264	480	266	105	143	Derby
404	967	433	275	343	447	426	554	272	315	208	570	459	465	378	240	777	451	301	433	Dover
362	256	320	473	417	338	336	172	586	426	549	901	790	288	378	669	214	616	491	299	Edinburgh
484	970	431	301	383	385	389	595	446	333	224	179	68	428	399	179	747	257	217	467	Exeter
459	851	369	333	394	257	304	525	533	338	333	550	439	322	344	372	620	114	249	407	Fishguard
583	106	518	668	623	520	518	414	795	631	742	1083	972	470	566	837	320	798	673	520	Fort William
402	283	338	484	439	340	338	230	612	447	547	900	789	290	386	668	137	615	489	340	Glasgow
291	906	344	216	245	402	354	478	101	241	216	568	457	425	306	241	698	429	293	359	Harwich
349	745	262	295	335	151	198	418	481	277	325	639	528	201	262	455	517	304	245	301	Holyhead
	615	90	145	61	196	151	190	240	148	262	663	552	180	103	412	425	433	272	63	Hull
382		579	729	673	605	604	428	842	682	816	1159	1049	555	634	925	417	872	747	552	Inverness
56	358		158	108	119	64	148	277	116	272	613	502	90	58	378	360	369	220	39	Leeds
90	453	98		82	187	138	301	187	42	119	480	369	206	105	225	510	282	109	174	Leicester
38	418	67	51		192	137	245	169	56	201	562	451	185	72	307	465	340	192	124	Lincoln
122	376	74	116	119		55	253	360	167	251	563	452	50	119	357	362	270	169	158	Liverpool
94	375	40	86	85	34		212	306	119	230	568	457	48	64	325	360	295	164	103	Manchester
118	266	92	187	152	157	132		414	254	412	774	663	203	206	518	262	517	378	127	Newcastle
149	523	174	116	105	224	190	257		198	224	626	515	354	241	299	637	439	283	293	Norwich
92	424	72	26	35	104	74	158	123		156	512	401	167	64	262	473	283	122	134	Nottingham
163	507	169	74	125	156	143	256	139	97		402	291	272	217	106	587	230	95	291	Oxford
412	720	381	298	349	350	353	481	389	318	250		126	597	560	357	925	436	396	645	Penzance
343	652	312	229	280	281	284	412	320	249	181	78		486	447	246	814	325	285	534	Plymouth
112	345	56	128	115	31	30	126	220	104	169	371	302		113	375	312	322	203	126	Preston
64	394	36	65	45	74	40	128	150	40	135	348	278	70		323	490	336	163	90	Sheffield
256	575	235	140	191	222	202	322	186	163	66	222	153	233	201		679	257	193	398	Southampton
264	259	224	317	289	225	224	163	396	294	365	575	506	194	254	422		641	515	362	Stranraer
269	542	229	175	211	168	183	321	273	176	143	271	202	200	209	160	398		161	418	Swansea
169	464	137	68	119	105	102	235	176	76	59	246	177	126	101	120	320	100		257	Worcester
39	343	24	108	77	98	64	79	182	83	181	401	332	78	56	247	225	260	160		York

Date	Action

Date	Action

Date	Action

Date	Action

Date	Action

Date	Action

Date	Action

Date	Action

Maestro & Montego Manuals

Like this handbook, the above manuals have been written specifically for your Maestro or Montego. Most models are covered from 1983 to 1991, the exceptions being MG Maestro 2.0 litre models, Maestro Van and Diesel engine models. The manuals are continually being updated – check with your local stockist to ensure that the latest-available edition is obtained.

The manuals are written in the same easy-to-follow manner as the handbooks and enable you to tackle even major overhauls and repairs yourself. They will advise you on which jobs are suitable for the DIY mechanic and often suggest practical alternatives to the specialist tools that are so often recommended.

With the ever increasing cost of garage labour, think of the money you could save!

Haynes Manuals are available from most good motor accessory stores and bookshops or direct from the publisher.

HAYNES

Haynes Publishing Group, Sparkford, Nr Yeovil, Somerset BA22 7JJ Telephone 0963 40635

AUSTIN/MG MAESTRO & MONTEGO